CalcLabs with Maple Commands
for Multivariable Calculus

Philip B. Yasskin
Art Belmonte
Texas A&M University

Prepared by

Philip B. Yasskin
Art Belmonte
Texas A&M University

BROOKS/COLE
CENGAGE Learning

Australia • Brazil • Japan • Korea • Mexico • Singapore • Spain • United Kingdom • United States

BROOKS/COLE
CENGAGE Learning

ISBN-13: 978-0-495-56058-6
ISBN-10: 0-495-56058-8

Brooks/Cole
10 Davis Drive
Belmont, CA 94002-3098
USA

Cengage Learning is a leading provider of customized learning solutions with office locations around the globe, including Singapore, the United Kingdom, Australia, Mexico, Brazil, and Japan. Locate your local office at:
www.cengage.com/international

Cengage Learning products are represented in Canada by Nelson Education, Ltd.

To learn more about Brooks/Cole, visit
www.cengage.com/brookscole

Purchase any of our products at your local college store or at our preferred online store
www.ichapters.com

Printed in the United States of America
1 2 3 4 5 6 7 12 11 10 09

Contents

In memory of our fathers

Stanton M. Yasskin

Arthur P. Belmonte, Sr.

Introduction

This is not a book on multivariable calculus. It is a lab manual on how to use *Maple* to help with multivariable calculus problems. The order of the material is organized by computational topic and should be suitable for most texts on multivariable calculus. For a review of how to use *Maple* to help with single variable calculus problems, see the lab manual *CalcLabs with Maple Commands for Single Variable Calculus* by Yasskin et al.

This book is accompanied by a *Maple* package called `VecCalc` which can be installed on any computer running *Maple* 7 through 12. Appendix B contains instructions for obtaining, installing and using the package. After starting the `VecCalc` package, you may get help on any command by executing

```
> ?VecCalc
```

and following the hyperlinks. Everything in this book has been tested with `VecCalc` 8.0 in *Maple* 10 through 12 using both the Classic Interface and the Worksheet Mode of the Standard Interface. Most of it should also works with *Maple* 7, 8, 9 or 9.5, except that a bug in *Maple* 9.5 causes Vectors of functions to display incorrectly (although they still operate correctly).

The book has two parts. The first part (chapters 1 through 8) explain how *Maple* can help with standard vector calculus computations. Each chapter ends with a set of short homework problems. The second part (chapters 9 and 10) contains assignments which could be used in a computer lab setting. Chapter 9 has shorter (one week) lab assignments, while chapter 10 has longer (multi-week) lab projects.

Chapter 1 covers the geometry of $\mathbb{R}^2$ and $\mathbb{R}^3$ including the algebra of vectors, the standard coordinate systems and the description of curves and surfaces.

Chapter 2 studies vector valued functions of one variable with emphasis on the properties of curves.

Chapter 3 discusses partial derivatives of functions of several variables and applications using tangent planes and directional derivatives.

Chapter 4 shows how *Maple* can automate the solution of max-min problems with several variables without or with constraints. The discussion is not restricted to just two variables.

Chapter 5 explains the commands for computing multiple integrals in rectangular, polar, cylindrical, spherical and general curvilinear coordinates. Applications include mass, center of mass and moment of inertia.

Chapter 6 studies how to compute line integrals and surface integrals using parametric curves and surfaces. Applications include mass, center of mass, moment of inertia, work, circulation, flux and expansion.

Chapter 7 discusses the commands for computing the gradient, divergence, curl, Laplacian and Hessian and how to find scalar and vector potentials.

Chapter 8 studies the major theorems of vector analysis: the Fundamental Theorem of Calculus for Curves, Green's Theorem, Stokes' Theorem and Gauss' Theorem. Applications include path and surface independence, work, circulation, flux, expansion and the computation of area and volume.

Chapter 9 is a collection of labs which might be used for one lab period in the computer lab. Typically the students work in pairs and have one week to complete the lab assignment. A short lab report is expected.

Chapter 10 is a collection of longer lab projects which require significant work. Typically the students work in groups of three or four and have two to four weeks to complete the project. An extensive project report is expected. Frequently, students need help from the instructor on such things as mathematical induction, making conjectures and proving them, writing *Maple* programs and how much documentation is expected in the *Maple* worksheet and the project report.

Appendix A discusses which Maple interface to use with this book, how to configure your worksheets and additional resources in the Standard Interface and at Maplesoft's website.

Appendix B contains instructions for obtaining, installing and using the `VecCalc` package.

Appendix C contains three tables which summarize the applications of integration which are computed throughout the book.

Appendix D is an introduction to *Maple* programming including conditional statements, looping structures and procedures or programs.

Revision History:

CalcLabs with Maple for Stewart's Multivariable Calculus, Concepts and Contexts by Belmonte and Yasskin, Brooks/Cole, 1998. ISBN 0-534-35739-3.
Multivariable CalcLabs with Maple for Stewart's Calculus, Fourth Edition by Belmonte and Yasskin, Brooks/Cole, 1999. ISBN 0-534-36444-6.
CalcLabs with Maple for Stewart's Multivariable Calculus, Concepts and Contexts, Second Edition by Belmonte and Yasskin, Brooks/Cole, 2001. ISBN 0-534-37914-1.
CalcLabs with Maple for Stewart's Multivariable Calculus, Fifth Edition by Yasskin and Belmonte, Brooks/Cole, 2003. ISBN 0-534-39361-6.
CalcLabs with Maple for Stewart's Multivariable Calculus, Concepts and Contexts, Third Edition by Yasskin and Belmonte, Brooks/Cole, 2005. ISBN 0-534-41010-3.
CalcLabs with Maple for Stewart's Multivariable Calculus, Sixth Edition by Yasskin and Belmonte, Brooks/Cole, 2005. ISBN 978-0-495-01231-3.

All author royalties on copies of this manual sold to Texas A&M University students have been donated to an undergraduate scholarship fund for Texas A&M University students.

Please send all comments and corrections to:

Philip Yasskin, Department of Mathematics, Texas A&M University, College Station, TX 77843-3368 or *yasskin@math.tamu.edu*

Syllabus for Calculus III

During a semester in a standard course of Calculus 3, the students should be able to cover Chapters. 1 through 8 and parts of 9 and 10 of this text. This assumes that students have one hour per week of organized computer laboratory instruction together with two or three hours per week of outside access to a computer during a fourteen week semester. In class, the students work in a self-paced mode, alone or in groups, using the computer to try the examples in the text and solve assigned exercises, labs or projects. The lab instructor should circulate to answer questions or provide guidance on troublesome syntax issues. Outside class, students should read the assigned chapters to prepare for lab and must complete the assigned exercises, labs or projects by the due date.

For each weekly lab, some instructors prefer to assign 4 or 5 exercises from the end of Chapters 1 – 8. Others assign one of the labs from Chapter 9. If the students do not complete the assignment during the lab, they can complete it at home and turn it in at the beginning of the next lab. For these weekly labs, students should work in pairs. Twice during the semester, the students can be assigned a project from Chapter 10, one on differential calculus and one on integral calculus. On the projects, students should work in groups of 3 or 4 and have from 2 to 4 weeks to complete it. Some instructors prefer to assign the same projects to the whole class. Other instructors assign different projects to each group.

Chapter 1

The Geometry of $\mathbb{R}^n$

1.1 Vector Algebra

Each time you start *Maple* and before you begin each section of this book, be sure you restart the `VecCalc` package as follows:

```
> with(VecCalc): VCalias:
```

When you load the `VecCalc` package, it automatically loads the `student` and `plots` packages. So you do not need to do that separately. For more details, see Appendix B. If you choose (as explained in footnote 1 below and in Appendix B) to work with `list`'s instead of `Vector` and `Matrix` types, then you should also execute:

```
> OutputVectorType:=list: OutputMatrixType:=list:
```

1.1.1 Scalars, Points and Vectors

In this book, a scalar is entered into *Maple* as a number or an expression and can be assigned a *name*. For example,

```
> a:=5; b:=x^2+1;
```

$$a := 5$$
$$b := x^2 + 1$$

Notice there are multiple *Maple* commands on a single line, each ending with a semi-colon (`;`). The symbol `:=` is called an assignment. The quantity on the right is "assigned" to a memory location whose name is given on the left. So the expression $x^2 + 1$ has been stored in a memory location and named `b`.

Similarly, a point or a vector (which is a sequence of 2 or more numbers or expressions) is entered using square brackets, `[ ]` or angle brackets, `< >`, (i.e. less than and greater than signs). For example,

```
> P:=<1|3|2>; u:=[p,q,r]; v:=<3,-4>;
```

$$P := [1,\ 3,\ 2]$$
$$u := [p,\ q,\ r]$$
$$v := \begin{bmatrix} 3 \\ -4 \end{bmatrix}$$

A sequence in square brackets is called a `list`; a sequence in angle brackets is called a `Vector`. In addition, if the components of a `Vector` are separated by vertical bars, the vector is displayed as a row, whereas, if the components are separated by commas, the vector is displayed as a column. In principle, mathematicians usually regard vectors as columns, for reasons which only become clear in future courses. However, to save space in this book, we will enter vectors as rows, using bars. You are free to enter a vector as a `list` or as either a row or column `Vector`.[1]

To display (or use) a scalar or vector, type its name.

```
> c:=2*b; v;
```

$$c := 2\,x^2 + 2$$
$$\begin{bmatrix} 3 \\ -4 \end{bmatrix}$$

To display (or use) a component of a vector, type its name followed by the component number in square brackets:

```
> v[2];
```

$$-4$$

Maple is not restricted to 2 or 3 dimensional vectors. (We will let $\mathbb{R}^2$ denote a 2-dimensional plane and let $\mathbb{R}^3$ denote 3-dimensional space.) *Maple* can handle vectors with any number of components. (We will let $\mathbb{R}^n$ denote n-dimensional space.) Further, the components do not need to be numbers. They can be undefined symbols, previously defined symbols or any expression using these:

```
> two_D:=<1|-6>; three_D:=<7|0|-4>; four_D:=<p|q|r|s>;
> <6|a|a*x^2-18|-8|45|w>;
```

$$two\_D := [1,\ -6]$$
$$three\_D := [7,\ 0,\ -4]$$
$$four\_D := [p,\ q,\ r,\ s]$$
$$[6,\ 5,\ 5\,x^2 - 18,\ -8,\ 45,\ w]$$

This last vector is an unnamed 6 dimensional vector. It contains the undefined variables `x` and `w` and a simple polynomial expression in `x`. Further, the previously defined variable `a` has been given its value of 5. If you don't want `a` to have its previous value, then you must first unassign it by typing

```
> a:='a';
```

[1] All of the commands in the `VecCalc` package work equally well if the input is a `list` or a `Vector` (or a `Matrix` to be discussed later). This book uses `Vector[row]` and `Matrix`, but some commands are easier with `list`. Normally the output type will match the input type. However, a few commands need to be told which type to use for output. To change the default output type, execute:

```
> OutputVectorType:=list: OutputMatrixType:=list:
```

The other choices for `OutputVectorType` are `Vector[row]` (the initial default) or `Vector[column]`. The other choice for `OutputMatrixType` is `Matrix` (the initial default).

$$a := a$$

Then we have

```
> <6|a|a*x^2-18|-8|45|w>;
```

$$[6,\ a,\ a\,x^2 - 18,\ -8,\ 45,\ w]$$

where a is undefined.

To compute the length of a vector,use the VecCalc command Length:

```
> v; length_of_v:=Length(v);
```

$$\begin{bmatrix} 3 \\ -4 \end{bmatrix}$$

$$length\_of\_v := 5$$

(If you did not get this result, it is probably because you did not load the VecCalc *package. Load it now, as explained in Appendix A.)*

This was easy and could have been done in your head, but consider:

```
> w:=<37/6|-41/28>; length_of_w:=Length(w);
```

$$w := \left[\frac{37}{6},\ \frac{-41}{28}\right]$$

$$length\_of\_w := \frac{\sqrt{283453}}{84}$$

1.1.2 Addition, Scalar Multiplication and Simplification

You can make more than one assignment in a single command by separating the variables and values by commas:

```
> u,v:=<1|-3|3>,<1|t|t^2>;
```

$$u,\ v := [1,\ -3,\ 3],\ [1,\ t,\ t^2]$$

You can add and subtract vectors and also multiply and divide a vector by a scalar by simply using the standard $+$, $-$, $*$ and $/$ signs:

```
> u+v, v-u, sqrt(2)*u, v/2;
```

$$[2,\ -3+t,\ 3+t^2],\ [0,\ t+3,\ t^2-3],\ \left[\sqrt{2},\ -3\sqrt{2},\ 3\sqrt{2}\right],\ \left[\frac{1}{2},\ \frac{t}{2},\ \frac{t^2}{2}\right]$$

The scalars and components of the vectors can even be expressions:

```
> 2*t*u+1/(1+2*t)*v;
```

$$\left[2\,t + \frac{1}{1+2\,t},\ -6\,t + \frac{t}{1+2\,t},\ 6\,t + \frac{t^2}{1+2\,t}\right]$$

To simplify this expression, click at the end of the previous line and add the simplify command:

```
> 2*t*u+1/(1+2*t)*v; simplify(%);
```

$$\left[2\,t + \frac{1}{1+2\,t},\ -6\,t + \frac{t}{1+2\,t},\ 6\,t + \frac{t^2}{1+2\,t}\right]$$

$$\left[\frac{2\,t+4\,t^2+1}{1+2\,t},\ -\frac{t\,(5+12\,t)}{1+2\,t},\ \frac{t\,(6+13\,t)}{1+2\,t}\right]$$

Here the percent sign (%) is *Maple*'s way of referring to the result of the immediately preceding computation. Note you can also do this with a single command:

```
> simplify(2*t*u+1/(1+2*t)*v);
```

$$\left[\frac{2\,t+4\,t^2+1}{1+2\,t},\ -\frac{t\,(5+12\,t)}{1+2\,t},\ \frac{t\,(6+13\,t)}{1+2\,t}\right]$$

The benefit of using two commands is that you can see the quantity to be computed before doing the operations. This prevents many mistakes due to typographical errors. There will be many more examples of this preventative measure later.

Example: 1.1. Find the distance between the points $P = (3, -2, 1)$ and $Q = (5, -3, 3)$.

SOLUTION: The vector from P to Q is the difference between the final point and the initial point: $\overrightarrow{PQ} = Q - P$. In *Maple* we compute

```
> P:=<3|-2|1>; Q:=<5|-3|3>; PQ:=Q-P;
```

$$P := [3,\ -2,\ 1]$$

$$Q := [5,\ -3,\ 3]$$

$$PQ := [2,\ -1,\ 2]$$

The distance from P to Q is then the length of this vector:

```
> distance_P_Q:=Length(PQ);
```

$$distance\_P\_Q := 3$$

The vector $\hat{v} = \dfrac{\vec{v}}{|\vec{v}|}$ is called the unit vector in the direction of $\vec{v}$ or simply the direction of $\vec{v}$. Throughout this book, a caret (^) over a vector indicates that it is a unit vector.

Example: 1.2. Find the unit vector in the direction of the vector $\vec{w} = \left\langle \dfrac{37}{6}, -\dfrac{41}{28} \right\rangle$. Give the exact answer and a decimal approximation.

SOLUTION: We define the vector $\vec{w}$ and compute the vector $\hat{w} = \dfrac{\vec{w}}{|\vec{w}|}$:

```
> w:=<37/6|-41/28>; w/Length(w); evalf(%);
```

$$w := \left[\frac{37}{6},\ \frac{-41}{28}\right]$$

$$\left[\frac{518\,\sqrt{283453}}{283453},\ -\frac{123\,\sqrt{283453}}{283453}\right]$$

$$[0.9729471067,\ -0.2310279809]$$

NOTE: The command `evalf(%)` forces *Maple* to evaluate the previous quantity as a floating point decimal.

1.1.3 The Dot Product

Recall that in any dimension the dot product of two vectors $\vec{u}$ and $\vec{v}$ is the sum of the products of corresponding components. For example, in $\mathbb{R}^3$ the dot product of $\vec{u}$ and $\vec{v}$ is:

$$\vec{u} \cdot \vec{v} = u_1 v_1 + u_2 v_2 + u_3 v_3 \ .$$

In *Maple* we can use the `VecCalc` command `Dot`. For example:

```
> u,v:=<2|5|-1>,<p|q|r>; Dot(u,v);
```

$$u,\ v := [2,\ 5,\ -1],\ [p,\ q,\ r]$$
$$2\,p + 5\,q - r$$

Alternatively, you can use the `VecCalc` operator `&.`:

```
> u &.  v;
```

$$2\,p + 5\,q - r$$

Further, in any dimension if you know the angle θ between two vectors $\vec{u}$ and $\vec{v}$, then their dot product may also be computed from:

$$\vec{u} \cdot \vec{v} = |\vec{u}|\,|\vec{v}| \cos(\theta) \ .$$

This formula may be solved for $\cos(\theta)$, and used for computing the angle between two vectors:

$$\cos(\theta) = \frac{\vec{u} \cdot \vec{v}}{|\vec{u}|\,|\vec{v}|} \ .$$

Recall that the `VecCalc` command `Length` will compute the length of a vector.

Example: 1.3. Space, the Final Frontier: As our navigator through the solar system, you notice that the Earth, Moon and Sun currently form a triangle with a 74.1° angle at the Earth. Find the angle θ (to the nearest hundredth of a degree) of the vertex at the Sun given that the distance from the Earth to the Sun is 390 times the distance from the Earth to the Moon. (The angles are in degrees for the primitive Earthlings.)

SOLUTION: Let a be the distance from the earth to the moon. Pick the coordinate system so that the earth is at the origin, $E = (0, 0)$, the sun is at $S = (390a, 0)$ and the moon is at $M = (a\cos(\theta), a\sin(\theta))$ where $\theta = 74.1°$. Since *Maple* computes all trig functions using radian measure, we first convert 74.1° into radians by using the `VecCalc` command `deg2rad` (or its alias `d2r`):

```
> theta:=d2r(74.1);
```

$$\theta := 1.293288976$$

Next we tell *Maple* that a is positive and enter the points S, E and M:

```
> assume(a>0);
> S:=<390*a|0>:  E:=<0|0>:  M:=<a*cos(theta)|a*sin(theta)>;
```

$$M := [0.2739592184\,a\tilde{}, 0.9617413096\,a\tilde{}]$$

NOTE: To save space in this book, we will sometimes omit the output of *Maple* commands when it is identical to the input, as for S and E above. This is done by using a colon (`:`) instead of a semi-colon (`;`) at the end of the statement. As a student, you should print out everything by using semi-colons to be sure the command is correct.

We then compute the vectors from S to E and from S to M and the length of these vectors:

```
> SE:=E-S; SM:=M-S;
```

$$SE := [-390\,a\tilde{}, 0]$$
$$SM := [-389.7260408\,a\tilde{}, 0.9617413096\,a\tilde{}]$$

```
> len_SE:=Length(SE); len_SM:=Length(SM);
```

$$len\_SE := 390\,a\tilde{}$$
$$len\_SM := 389.7272274\,a\tilde{}$$

Next we compute $\cos(\theta)$:

```
> cos_theta:=Dot(SE,SM) / (len_SE*len_SM); #DON'T FORGET PARENTHESES
```

$$cos\_theta := 0.9999969553$$

NOTE: On a *Maple* input line, anything following a # is a comment which *Maple* ignores.

Finally, we take the arccos to get θ:

```
> theta:=arccos(cos_theta);
```

$$\theta := 0.002467671593$$

The `VecCalc` package also contains a command `Angle` which computes this angle directly:

```
> theta:=Angle(SE,SM);
```

$$\theta := 0.002467671593$$

Since *Maple* computes all inverse trig functions using radian measure, this value for theta is in radians. To convert it to degrees you can use the `VecCalc` command `rad2deg` (or its alias `r2d`):

```
> r2d(theta);
```

$$0.1413871675$$

Thus, to the nearest hundredth of a degree, the angle is $\theta = 0.14$ degrees.

Another application of the dot product is to compute the scalar and vector projections of a vector $\vec{u}$ along a vector $\vec{v}$ and the orthogonal projection of $\vec{u}$ perpendicular to $\vec{v}$. These are shown in the Figure.

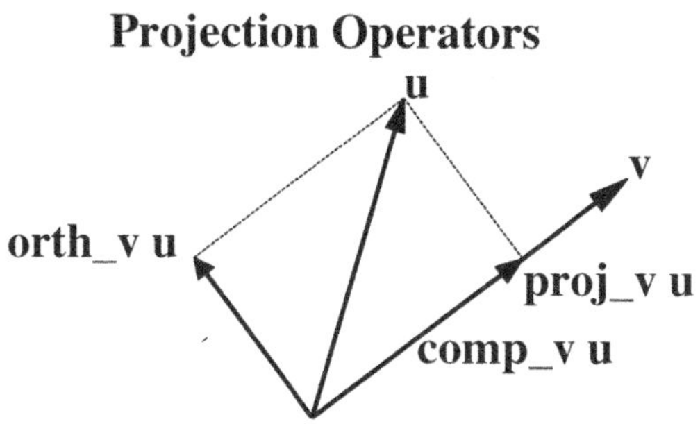

The scalar projection or component of $\vec{u}$ along $\vec{v}$ is computed from the formula:

$$\text{comp}_{\vec{v}}\vec{u} = \frac{\vec{u}\cdot\vec{v}}{|\vec{v}|} = \vec{u}\cdot\hat{v},$$

where $\hat{v} = \dfrac{\vec{v}}{|\vec{v}|}$. The vector projection of $\vec{u}$ along $\vec{v}$ is computed from the formula:

$$\text{proj}_{\vec{v}}\vec{u} = \frac{\vec{u}\cdot\vec{v}}{|\vec{v}|^2}\vec{v} = (\vec{u}\cdot\hat{v})\hat{v}$$

The projection of $\vec{u}$ orthogonal to $\vec{v}$ is computed from the formula:

$$\text{proj}_{\perp\vec{v}}\vec{u} = \vec{u} - \text{proj}_{\vec{v}}\vec{u} = \vec{u} - \frac{\vec{u}\cdot\vec{v}}{|\vec{v}|^2}\vec{v} = \vec{u} - (\vec{u}\cdot\hat{v})\hat{v}$$

Example: 1.4. For the vectors $\vec{a} = -\hat{\imath} - 2\hat{\jmath} + 2\hat{k}$ and $\vec{b} = 3\hat{\imath} + 3\hat{\jmath} + 4\hat{k}$, find the scalar and vector projections of $\vec{b}$ along $\vec{a}$ and the projection of $\vec{b}$ orthogonal to $\vec{a}$.

SOLUTION: Define the vectors:

```
> a:=<-1|-2|2>:  b:=<3|3|4>:
```

Compute the scalar projection:

```
> scal_proj:=Dot(b,a)/Length(a);
```

$$scal\_proj := \frac{-1}{3}$$

Compute the vector projection:

```
> vect_proj:=Dot(b,a)/Length(a)^2 * a;
```

$$vect\_proj := \left[\frac{1}{9}, \frac{2}{9}, \frac{-2}{9}\right]$$

Compute the orthogonal projection:

```
> orthog_proj:=b - vect_proj;
```

$$orthog\_proj := \left[\frac{26}{9}, \frac{25}{9}, \frac{38}{9}\right]$$

Example: 1.5. Compute the work done on a box by a horizontal force of 35 lbs which moves the box 9 ft up a ramp which is inclined at an angle of 15 degrees.

SOLUTION: We input the force and distance and convert the 15° angle into radian measure by using the `VecCalc` command `d2r`:

```
> F:=35: d:=9: theta:=d2r(15);
```

$$\theta := \frac{\pi}{12}$$

The work done is the dot product of the force vector and the displacement vector. Since we know the magnitude of these vectors and the angle between them, we use the angle formula for the dot product:

```
> work:=F*d*cos(theta); convert(%,'radical'); evalf(%);
```

$$work := 315\cos(\frac{\pi}{12})$$

$$\frac{315\sqrt{2}\,(1+\sqrt{3})}{4}$$

$$304.2666353$$

So the work done is 304.3 ft-lbs. (Telling the boss the work done is $\dfrac{315\sqrt{2}\,(1+\sqrt{3})}{4}$ ft-lb is a good way to get fired.)

1.1.4 The Cross Product

The cross product can only be defined in 3 dimensions. Given two vectors $\vec{u} = (u_1, u_2, u_3)$ and $\vec{v} = (v_1, v_2, v_3)$, their cross product is defined to be the vector

$$\vec{u} \times \vec{v} = \begin{vmatrix} \hat{\imath} & \hat{\jmath} & \hat{k} \\ u_1 & u_2 & u_3 \\ v_1 & v_2 & v_3 \end{vmatrix} = (u_2v_3 - u_3v_2)\hat{\imath} + (u_3v_1 - u_1v_3)\hat{\jmath} + (u_1v_2 - u_2v_1)\hat{k}$$

$$= (u_2v_3 - u_3v_2,\ u_3v_1 - u_1v_3,\ u_1v_2 - u_2v_1)$$

In *Maple* you can compute the cross product by using the `VecCalc` command `Cross`:

```
> u:=<2|5|-1>: v:=<p|q|r>: Cross(u,v);
```

$$[5\,r + q,\ -p - 2\,r,\ 2\,q - 5\,p]$$

Alternatively, you can use the `VecCalc` operator `&x`:

```
> u &x v;
```

$$[5\,r + q,\ -p - 2\,r,\ 2\,q - 5\,p]$$

NOTE: You need a space after the `&x` so that *Maple* will know that the `v` is not part of the operator.

Example: 1.6. If $\vec{a} = (-2, 3, 4)$ and $\vec{b} = (3, 0, 1)$, compute $\vec{a} \times \vec{b}$.

SOLUTION: Enter the vectors and compute the cross product.

```
> a:=<-2|3|4>:  b:=<3|0|1>:  axb:=Cross(a,b);
```

$$axb := [3,\ 14,\ -9]$$

As applications of the cross product, we have:

1. The area of a parallelogram with edges $\vec{u}$ and $\vec{v}$ is the length of their cross product:

$$A_{\text{para}} = |\vec{u} \times \vec{v}|$$

2. The area of a triangle with edges $\vec{u}$ and $\vec{v}$ is half of the length of their cross product:

$$A_{\text{tri}} = \frac{1}{2}\,|\vec{u} \times \vec{v}|$$

3. The volume of a parallelepiped with edges $\vec{u}$, $\vec{v}$ and $\vec{w}$ is the absolute value of their triple product:

$$V_{\text{para}} = |(\vec{u} \times \vec{v}) \cdot \vec{w}|$$

Example: 1.7. Find the area of the triangle with vertices $P = (3, 2, -5)$, $Q = (0, -2, 3)$ and $R = (-5, -1, 2)$.

SOLUTION: Enter the points and compute two edge vectors:

```
> P:=<3|2|-5>:  Q:=<0|-2|3>:  R:=<-5|-1|2>:
> PQ:=Q-P; PR:=R-P;
```

$$PQ := [-3,\ -4,\ 8]$$

$$PR := [-8,\ -3,\ 7]$$

Now compute the area as half of the length of the cross product.

```
> cp:=Cross(PQ,PR); area:=Length(cp)/2;
```

$$cp := [-4,\ -43,\ -23]$$

$$area := \frac{3\sqrt{266}}{2}$$

Example: 1.8. Find the volume of the parallelepiped with edges $\vec{a} = (0, 0, 1)$, $\vec{b} = (0, 2, 2)$ and $\vec{c} = (3, 3, 3)$.

SOLUTION: Enter the edge vectors, compute the triple product and its absolute value:

```
> a:=<0|0|1>:  b:=<0|2|2>:  c:=<3|3|3>:
> (a &x b) &.  c; V:=abs(%);
```

$$-6$$

$$V := 6$$

1.2 Coordinates

Remember to restart the `VecCalc` package.

1.2.1 Polar Coordinates in $\mathbb{R}^2$

In $\mathbb{R}^2$, there are two standard coordinate systems: a point P has rectangular coordinates (x, y) and polar coordinates (r, θ). These coordinates are shown in the Figure.

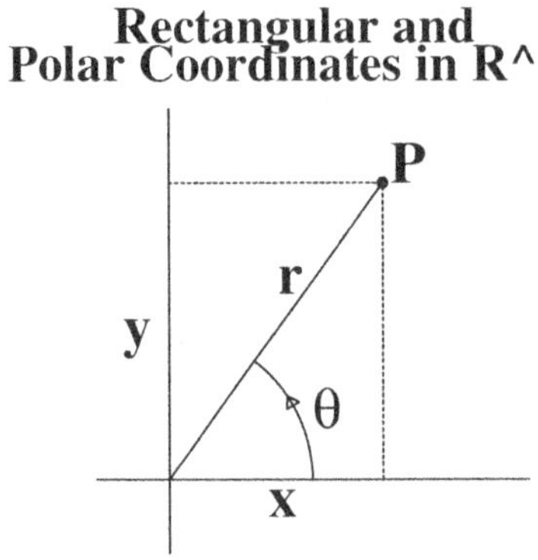

The `VecCalc` command `polar2rect` (or `p2r`) converts from polar to rectangular coordinates. The `VecCalc` command `rect2polar` (or `r2p`) converts from rectangular to polar coordinates. Both `p2r` and `r2p` expect a single argument which is a list or vector of two coordinates. If the argument of `p2r` or `r2p` contains any floating point decimal numbers, then `p2r` and `r2p` return decimal answers. Otherwise, they return exact numbers or symbolic expressions.

Here are some examples:

```
> p2r([r,theta]), p2r(<2,Pi/6>), simplify(p2r(<2.|Pi/6>));
```

$$[r\cos(\theta),\, r\sin(\theta)],\ \begin{bmatrix}\sqrt{3}\\ 1\end{bmatrix},\ [1.732050808,\ 1.]$$

```
> r2p(<x,y>), r2p([-2,0]), r2p([-2,-2]);
```

$$\begin{bmatrix}\sqrt{x^2+y^2}\\ \arctan(y,\, x)\end{bmatrix},\ [2,\ \pi],\ [2\sqrt{2},\ -\frac{3\pi}{4}]$$

```
> r2p([3,-4]), r2p(<3.|-4>);
```

$$[5,\ -\arctan(\frac{4}{3})],\ [5.,\ -0.9272952180]$$

NOTE: The *Maple* command `arctan(y,x)` with 2 arguments is precisely designed to produce exactly what is needed for θ:

```
> arctan(1,1), arctan(1,-1), arctan(-1,-1), arctan(-1,1);
```

$$\frac{\pi}{4},\ \frac{3\pi}{4},\ -\frac{3\pi}{4},\ -\frac{\pi}{4}$$

1.2.2 Cylindrical and Spherical Coordinates in $\mathbb{R}^3$

In $\mathbb{R}^3$, there are three standard coordinate systems: a point P has rectangular coordinates (x, y, z), cylindrical coordinates (r, θ, z) and spherical coordinates (ρ, θ, ϕ). These coordinates are shown in the Figure.

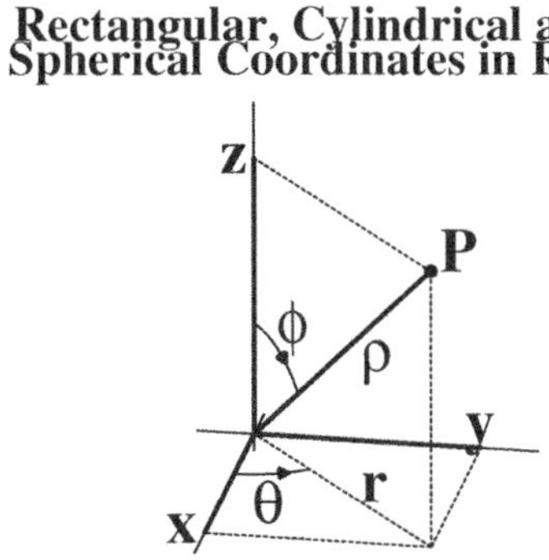

There are 6 `VecCalc` commands which convert between rectangular, cylindrical and spherical coordinates:

- `cyl2rect` (or `c2r`) converts from cylindrical to rectangular coordinates.
- `rect2cyl` (or `r2c`) converts from rectangular to cylindrical coordinates.
- `sph2rect` (or `s2r`) converts from spherical to rectangular coordinates.
- `rect2sph` (or `r2s`) converts from rectangular to spherical coordinates.
- `sph2cyl` (or `s2c`) converts from spherical to cylindrical coordinates.
- `cyl2sph` (or `c2s`) converts from cylindrical to spherical coordinates.

Each of these commands expects a single argument which is a list or vector of three coordinates. If the argument contains any floating point decimal numbers, then these commands return decimal answers. Otherwise, they return exact numbers or symbolic expressions.

Here are some examples:

```
> c2r([r,theta,z]), r2c([x,y,z]);
```

$$[r\cos(\theta),\, r\sin(\theta),\, z],\ [\sqrt{x^2+y^2},\, \arctan(y,\, x),\, z]$$

```
> s2r([rho,theta,phi]); r2s([x,y,z]);
```

$$[\rho\sin(\phi)\cos(\theta),\, \rho\sin(\phi)\sin(\theta),\, \rho\cos(\phi)]$$

$$[\sqrt{x^2+y^2+z^2},\, \arctan(y,\, x),\, \arctan(\sqrt{x^2+y^2},\, z)]$$

```
> s2c([rho,theta,phi]), c2s([r,theta,z]);
```

$$[\rho\sin(\phi),\, \theta,\, \rho\cos(\phi)],\ [\sqrt{r^2+z^2},\, \theta,\, \arctan(r,\, z)]$$

```
> c2r([2, -Pi/3,4]), r2c([3,4,12]);
```

$$[1, -\sqrt{3}, 4], [5, \arctan(\frac{4}{3}), 12]$$

```
> s2r([1,Pi/4,Pi/4]), r2s([.5,.5,1/sqrt(2)]);
```

$$[\frac{1}{2}, \frac{1}{2}, \frac{\sqrt{2}}{2}], [1., 0.7853981634, 0.7853981635]$$

```
> s2c([1,Pi/4,Pi/4]), c2s([5,theta,12]);
```

$$[\frac{\sqrt{2}}{2}, \frac{\pi}{4}, \frac{\sqrt{2}}{2}], [13, \theta, \arctan(\frac{5}{12})]$$

1.3 Curves and Surfaces

1.3.1 Lines and Planes

Parametric Lines To specify a line one can give either (i) two points P and Q on the line or (ii) one point P and a direction given by a vector $\vec{v}$ tangent to the line. Given two points on the line, the direction for the line can be taken as the vector between the two points $\vec{v} = \overrightarrow{PQ} = Q - P$. We want to find an equation for the general point X on the line.

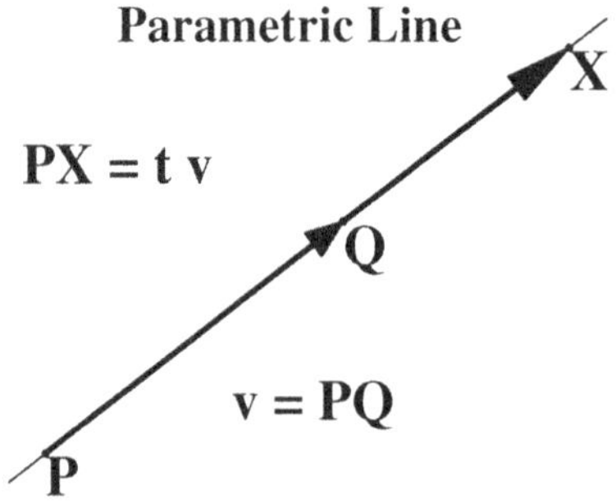

From the Figure, notice that the vector from P to X is a multiple of the vector $\vec{v}$. Letting t denote the proportionality constant, we have

$$\overrightarrow{PX} = t\vec{v} \qquad \text{or} \qquad X - P = t\vec{v} \qquad \text{or} \qquad X = P + t\vec{v}$$

These are the parametric equations for a line and t (called the parameter) says where you are on the line. The vector $\vec{v}$ is called a direction vector or a tangent vector for the line.

Example: 1.9. Find parametric equations for the line through the points $P = (2, -1, 3)$ and $Q = (5, 2, 4)$.

SOLUTION: We define the points and the direction vector:

```
> P:=<2|-1|3>: Q:=<5|2|4>: v:=Q-P;
```

$$v := [3, 3, 1]$$

We define $X = (x, y, z)$ as the generic point and construct the equation of the line:

```
> X:=<x|y|z>:  line1:=X=simplify(P+t*v);
```

$$line1 := [x,\ y,\ z] = [2+3t,\ -1+3t,\ 3+t]$$

To write this as separate equations, we use the `equate` command from the `student` package:

```
> line2:=equate(X,simplify(P+t*v));
```

$$line2 := \{x = 2+3t,\ y = -1+3t,\ z = 3+t\}$$

NOTE: The `student` package is automatically loaded when you load the `VecCalc` package.

Parametric Planes Similarly, to specify a plane one can give either (i) three points P, Q and R on the plane or (ii) one point P and two vectors $\vec{u}$ and $\vec{v}$ tangent to the plane or (iii) one point P and one vector $\vec{N}$ (called the normal vector) perpendicular to the plane. Given three points, the two tangent vectors can be taken as $\vec{u} = \overrightarrow{PQ} = Q - P$ and $\vec{v} = \overrightarrow{PR} = R - P$. (See the Figure "Parametric Plane" below.) Given the two vectors, the normal vector can be taken as $\vec{N} = \vec{u} \times \vec{v}$. (See the Figure "Non-Parametric Plane" further below.) We want to find an equation for the general point X on the plane.

Parametric Plane

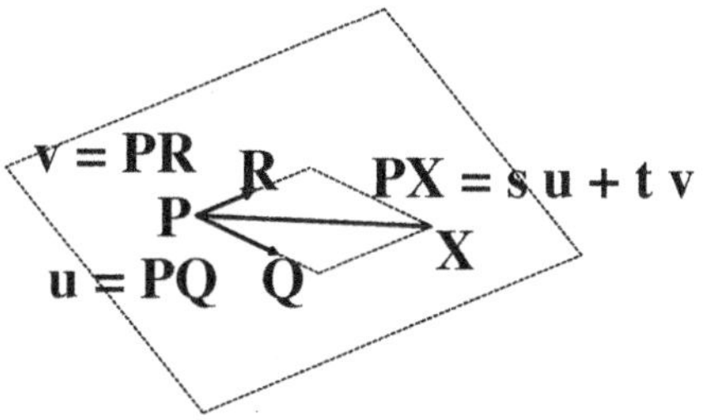

Given a point and two tangent vectors, notice that the vector from P to X can be written as a multiple of the vector $\vec{u}$ plus a multiple of the vector $\vec{v}$. See the Figure. Letting s and t be the multiples, we have

$$\overrightarrow{PX} = s\vec{u} + t\vec{v} \qquad \text{or} \qquad X - P = s\vec{u} + t\vec{v} \qquad \text{or} \qquad X = P + s\vec{u} + t\vec{v}$$

These are the parametric equations for a plane and s and t (called the parameters) determine where you are on the plane.

Example: 1.10. Find parametric equations for the plane through the points $P = (2, -1, 3)$, $Q = (5, 2, 4)$ and $R = (-4, 2, 2)$.

SOLUTION: We define the points and the two vectors between them:

```
> P:=<2|-1|3>:  Q:=<5|2|4>:  R:=<-4|2|2>:
> u:=Q-P;    v:=R-P;
```

$$u := [3,\ 3,\ 1]$$
$$v := [-6,\ 3,\ -1]$$

We define $X = (x, y, z)$ as the generic point and construct the equation of the plane:

```
> X:=<x|y|z>:  plane1:=X=simplify(P+s*u+t*v);
```

$$plane1 := [x,\ y,\ z] = [2 + 3\,s - 6\,t,\ -1 + 3\,s + 3\,t,\ 3 + s - t]$$

To write this as separate equations, we use `equate`:

```
> plane2:=equate(X,simplify(P+s*u+t*v));
```

$$plane2 := \{x = 2 + 3\,s - 6\,t,\ y = -1 + 3\,s + 3\,t,\ z = 3 + s - t\}$$

Non-Parametric Planes Alternatively, given a point and a normal vector, notice that the vector from P to X is perpendicular to $\vec{N}$. See the Figure.

Non-Parametric Plane

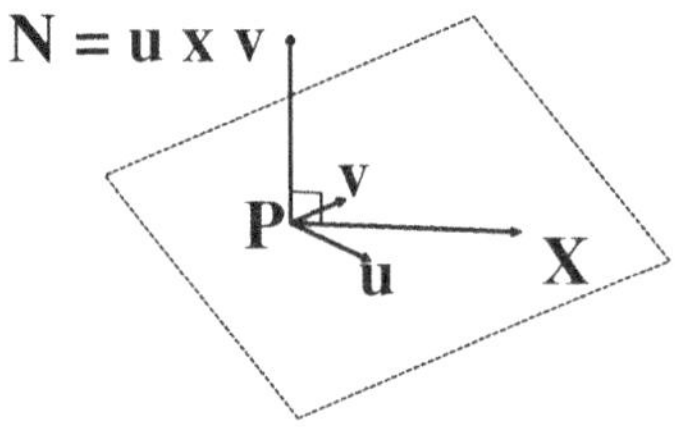

Thus:

$$\vec{N} \cdot \overrightarrow{PX} = 0 \qquad \text{or} \qquad \vec{N} \cdot (X - P) = 0 \qquad \text{or} \qquad \vec{N} \cdot X = \vec{N} \cdot P$$

This is a (non-parametric) equation for the plane.

Example: 1.11. Find the non-parametric equation for the plane through the points $P = (2, -1, 3)$, $Q = (5, 2, 4)$ and $R = (-4, 2, 2)$.

SOLUTION: We define the points and two vectors as above and then construct the normal vector:

```
> N:=Cross(u,v);
```

$$N := [-6,\ -3,\ 27]$$

We enter the generic point, $X = (x, y, z)$, and find the equation of the plane:

```
> X:=<x|y|z>:  plane3:=Dot(N,X) = Dot(N,P);
```

$$plane3 := -6\,x - 3\,y + 27\,z = 72$$

Finally, notice that this is equivalent to the equation which is obtained by eliminating the parameters in the parametric equations:

```
> solve({plane2[1],plane2[2]},{s,t});
```

$$\{s = \frac{x}{9} + \frac{2\,y}{9},\ t = \frac{y}{9} + \frac{1}{3} - \frac{x}{9}\}$$

```
> subs(%,plane2[3]);  27*%;
```

$$z = \frac{8}{3} + \frac{2\,x}{9} + \frac{y}{9}$$

$$27\,z = 72 + 6\,x + 3\,y$$

Non-Parametric Lines So far we have discussed parametric lines and planes and non-parametric planes. It remains to discuss non-parametric lines. The situation is different in $\mathbb{R}^2$ and $\mathbb{R}^3$.

In $\mathbb{R}^2$, the non-parametric equations for the line through a point P with normal vector $\vec{n}$ is given by: (See the Figure.)

$$\vec{n} \cdot \overrightarrow{PX} = 0 \qquad \text{or} \qquad \vec{n} \cdot (X - P) = 0 \qquad \text{or} \qquad \vec{n} \cdot X = \vec{n} \cdot P$$

If a direction vector for the line is $\vec{v} = (v_1, v_2)$, the normal vector may be taken as $\vec{n} = (v_2, -v_1)$, since then $\vec{n} \cdot \vec{v} = 0$.

Non-Parametric Line in 2D

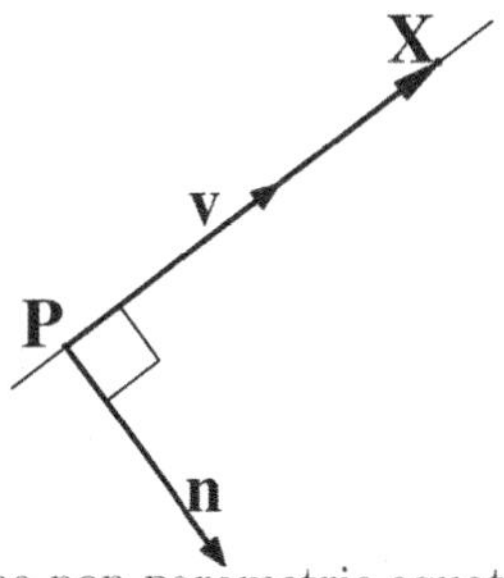

Example: 1.12. Find the non-parametric equations for the line through the points $A = (4, 7)$ and $B = (-2, 3)$.

SOLUTION: We enter the points and find the direction vector:

```
> A:=<4|7>: B:=<-2|3>: v:=B-A;
```

$$v := [-6,\ -4]$$

So the normal vector is

```
> n:=<v[2]|-v[1]>;
```

$$n := [-4,\ 6]$$

Then we enter the generic point, $X = (x, y)$, and find the equation of the line:

```
> X:=<x|y>: line2d:=Dot(n,X) = Dot(n,A);
```

$$line2d := -4\,x + 6\,y = 26$$

In $\mathbb{R}^3$, the non-parametric or symmetric equations for the line through the point $P = (p, q, r)$ with direction vector $\vec{v} = (a, b, c)$ are:

$$\frac{x-p}{a} = \frac{y-q}{b} = \frac{z-r}{c}.$$

This says the line is the intersection of the two planes:

$$\frac{x-p}{a} = \frac{y-q}{b} \quad \text{and} \quad \frac{y-q}{b} = \frac{z-r}{c}.$$

Example: 1.13. Find the symmetric equations for the line through the points $P = (2, -1, 3)$ and $Q = (5, 2, 4)$.

SOLUTION: We enter the points and find the direction vector:

```
> P:=<2|-1|3>:  Q:=<5|2|4>:  v:=Q-P;
```

$$v := [3,\ 3,\ 1]$$

Reading off coefficients, we construct the two equations for the line:

```
> line3:={(x-P[1])/v[1] = (y-P[2])/v[2], (y-P[2])/v[2] =
> (z-P[3])/v[3]};
```

$$line3 := \{\frac{x}{3} - \frac{2}{3} = \frac{y}{3} + \frac{1}{3}, \frac{y}{3} + \frac{1}{3} = z - 3\}$$

These are the equations of two planes whose intersection is the line.

1.3.2 Quadric Curves and Quadric Surfaces

Quadric Curves A quadric curve is the graph of a quadratic equation in $\mathbb{R}^2$. The general quadratic equation with no cross terms is $Ax^2 + By^2 + Cx + Dy + E = 0$. By completing the squares on x and y (when possible), it may be brought to one of the following standard forms:

$(x-p)^2 + (y-q)^2 = r^2$ .... circle

$\dfrac{(x-p)^2}{a^2} + \dfrac{(y-q)^2}{b^2} = 1$ .... ellipse

$\dfrac{(x-p)^2}{a^2} - \dfrac{(y-q)^2}{b^2} = \pm 1$ .... hyperbola

$\dfrac{(x-p)^2}{a^2} - \dfrac{(y-q)^2}{b^2} = 0$ .... cross

$y - q = \pm a(x-p)^2$ or $x - p = \pm a(y-q)^2$ .... parabola

Example: 1.14. Classify and plot the following quadric curves:

a) $4x^2 + 9y^2 - 16x + 18y = 11$

b) $4x^2 - 9y^2 - 16x - 18y = 29$

- For a circle, give the center and radius.
- For an ellipse, give the center and semi-radii.
- For a hyperbola, give the center, direction and asymptotes and add the asymptotes to the plot.
- For a cross, give the intersection point and the two lines.
- For a parabola, give the vertex and the direction.

SOLUTION: For each equation, we enter the equation as an expression, complete the squares using `completesquare` from the `student` package and manipulate the equation into a standard form. Then we classify the curve and plot the equation using the `implicitplot` command from the `plots` package.

NOTE: The `student` and `plots` packages are automatically loaded by the `VecCalc` package.

a) We enter the equation and complete the squares:

```
> eq1:=4*x^2 + 9*y^2 - 16*x + 18*y = 11:
> eq2:=completesquare(eq1,{x,y});
```

$$eq2 := 9\,(y+1)^2 - 25 + 4\,(x-2)^2 = 11$$

Maple knows how to add two equations and how to multiply or divide an equation by a number:

```
> eq3:=eq2 + (25=25);
```

$$eq3 := 9\,(y+1)^2 + 4\,(x-2)^2 = 36$$

```
> eq4:=eq3/36;
```

$$eq4 := \frac{(y+1)^2}{4} + \frac{(x-2)^2}{9} = 1$$

This is the standard equation for an ellipse with center at $(2,-1)$ and radii 3 in the x-direction and 2 in the y-direction. Its graph is:

```
> implicitplot(eq4, x=-5..5, y=-5..5, scaling=constrained);
```

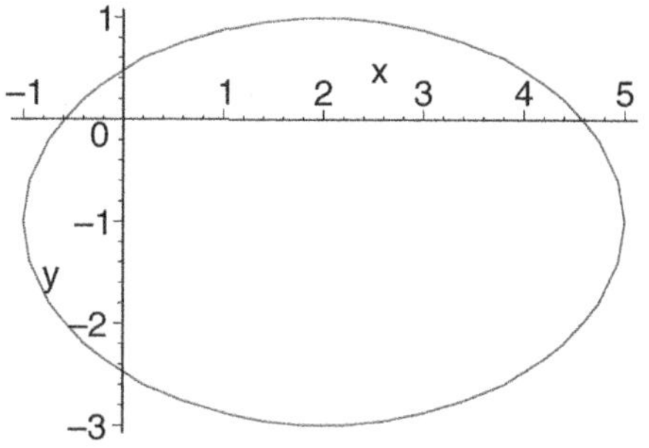

b) We enter the equation, complete the squares and manipulate it into a standard form:

```
> eq1:=4*x^2 - 9*y^2 - 16*x - 18*y = 29:
> eq2:=completesquare(eq1,{x,y});
```

$$eq2 := -9\,(y+1)^2 - 7 + 4\,(x-2)^2 = 29$$

```
> eq3:=eq2 + (7=7);
```

$$eq3 := -9\,(y+1)^2 + 4\,(x-2)^2 = 36$$

```
> eq4:=eq3/36;
```

$$eq4 := -\frac{(y+1)^2}{4} + \frac{(x-2)^2}{9} = 1$$

This is the standard equation for a hyperbola with center at $(2, -1)$ which opens along the positive and negative x-axis. Its asymptotes are the cross obtained by replacing the 1 on the right hand side by a 0:

NOTE: The commands `lhs` and `rhs` read off the left and right hand sides of an equation.

```
> asymptotes:=lhs(eq4)=0;
```

$$asymptotes := -\frac{(y+1)^2}{4} + \frac{(x-2)^2}{9} = 0$$

```
> solve(asymptotes,y);
```

$$-\frac{2x}{3} + \frac{1}{3}, \frac{2x}{3} - \frac{7}{3}$$

So the asymptotes are $y = -\frac{2}{3}x + \frac{1}{3}$ and $y = \frac{2}{3}x - \frac{7}{3}$. Finally, we plot the hyperbola and its asymptotes:

```
> implicitplot({eq4,asymptotes}, x=-5..9, y=-6..4,
> scaling=constrained, grid=[49,49]);
```

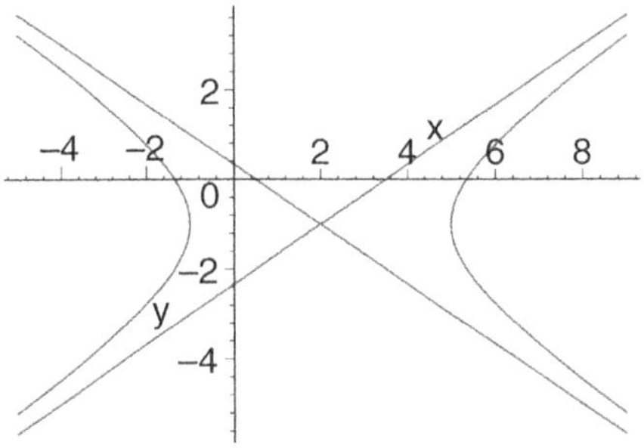

NOTE: The `grid` option specifies the number of points to use in each direction.

Quadric Surfaces A quadric surface is the graph of a quadratic equation in $\mathbb{R}^3$. The general quadratic equation with no cross terms is $Ax^2 + By^2 + Cz^2 + Dx + Ey + Fz + G = 0$. By completing the squares on x, y and z (when possible), it may be brought to one of the following standard forms (up to the rearrangement of x, y and z):

$(x-p)^2 + (y-q)^2 + (z-r)^2 = R^2$ .... sphere

$\frac{(x-p)^2}{a^2} + \frac{(y-q)^2}{b^2} + \frac{(z-r)^2}{c^2} = 1$ .... ellipsoid

$\frac{(x-p)^2}{a^2} + \frac{(y-q)^2}{b^2} - \frac{(z-r)^2}{c^2} = 1$ .... hyperboloid of 1 sheet

$-\frac{(x-p)^2}{a^2} - \frac{(y-q)^2}{b^2} + \frac{(z-r)^2}{c^2} = 1$ .... hyperboloid of 2 sheets

$\frac{(x-p)^2}{a^2} + \frac{(y-q)^2}{b^2} - \frac{(z-r)^2}{c^2} = 0$ .... cone

$z - r = \dfrac{(x-p)^2}{a^2} + \dfrac{(y-q)^2}{b^2}$ .... elliptic paraboloid

$z - r = \dfrac{(x-p)^2}{a^2} - \dfrac{(y-q)^2}{b^2}$ .... hyperbolic paraboloid

A quadratic equation in two coordinates .... cylinder whose cross section is the quadric curve

Example: 1.15. Classify and plot the following quadric surfaces:

a) $4x^2 - y^2 - 9z^2 - 16x - 2y + 18z = 30$

b) $4x^2 - 9z^2 - 16x + 2y - 18z = -1$

- For a sphere, give the center and radius.
- For an ellipsoid, give the center and semi-radii.
- For a hyperboloid, say whether it has 1 or 2 sheets, give the center, axis and asymptotic cone and plot the asymptotic cone.
- For a cone, give the vertex and direction.
- For a paraboloid, say whether it is elliptic or hyperbolic and give the vertex and the direction(s).
- For a cylinder, give its axis and its cross section.

SOLUTION: For each equation, we enter the equation as an expression, complete the squares using `completesquare` from the `student` package and manipulate the equation into a standard form. Then we classify the curve and plot the equation using the `implicitplot3d` command from the `plots` package.

a) We enter the equation, complete the squares and manipulate it into a standard form:

```
> eq1:=4*x^2 - y^2 - 9*z^2 - 16*x - 2*y + 18*z = 30:
> eq2:=completesquare(eq1,{x,y,z});
```

$$eq2 := -9\,(z-1)^2 - 6 - (y+1)^2 + 4\,(x-2)^2 = 30$$

```
> eq3:=eq2 + (6=6);
```

$$eq3 := -9\,(z-1)^2 - (y+1)^2 + 4\,(x-2)^2 = 36$$

```
> eq4:=eq3/36;
```

$$eq4 := -\frac{(z-1)^2}{4} - \frac{(y+1)^2}{36} + \frac{(x-2)^2}{9} = 1$$

This is the standard equation for a hyperboloid of 2 sheets with center at $(2, -1, 1)$ and axis which is parallel to the x-axis. Its asymptotic cone is obtained by replacing the 1 on the right hand side by a 0:

```
> asymptote:=lhs(eq4)=0;
```

$$asymptote := -\frac{(z-1)^2}{4} - \frac{(y+1)^2}{36} + \frac{(x-2)^2}{9} = 0$$

Finally, we plot the hyperboloid and the asymptotic cone.

```
> implicitplot3d(eq4, x=-6..10, y=-18..16, z=-6..8, axes=normal,
> grid=[15,15,15], scaling=constrained, orientation=[85,85]);
> implicitplot3d(asymptote, x=-6..10, y=-18..16, z=-6..8, axes=normal,
> grid=[15,15,15], scaling=constrained, orientation=[85,85]);
```

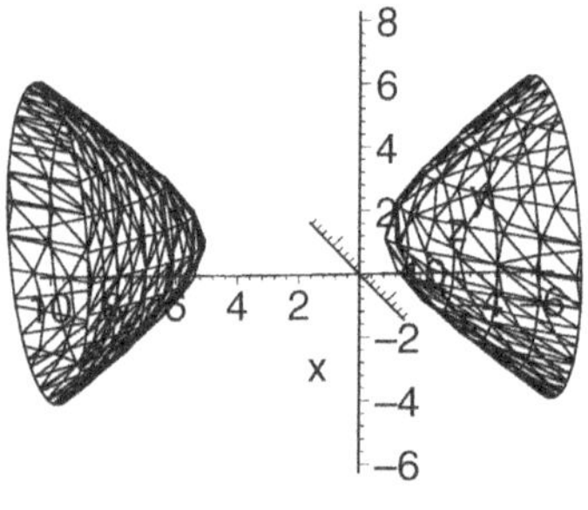

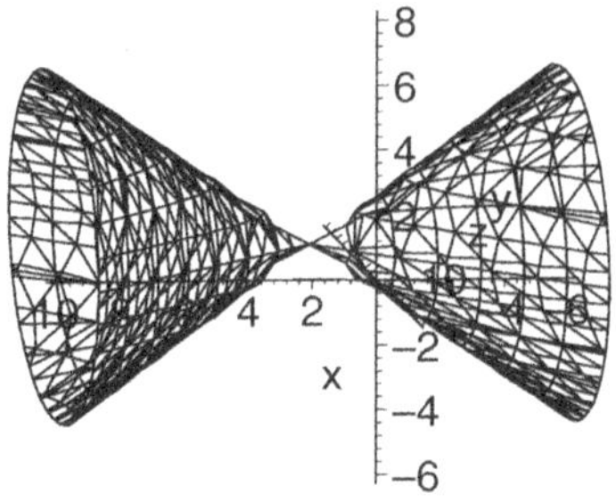

b) Since the equation is linear in y, we enter the equation, solve for y and then complete the squares:

```
> eq1:=4*x^2 - 9*z^2 - 16*x + 2*y - 18*z = -1:
> eq2:=y=solve(eq1, y);
```

$$eq2 := y = -2\,x^2 + \frac{9}{2}\,z^2 + 8\,x + 9\,z - \frac{1}{2}$$

```
> eq3:=completesquare(eq2,{x,z});
```

$$eq3 := y = \frac{9\,(z+1)^2}{2} + 3 - 2\,(x-2)^2$$

This is the standard equation for a hyperbolic paraboloid with vertex at $(2, 3, -1)$ and axis parallel to the y-axis, which opens upward in the zy-plane and downward in the xy-plane. Finally, we plot the hyperbolic paraboloid:

```
> implicitplot3d(eq3, x=-3..7, y=-3..8, z=-6..4, axes=normal,
> grid=[15,15,15], scaling=constrained, orientation=[35,75]);
```

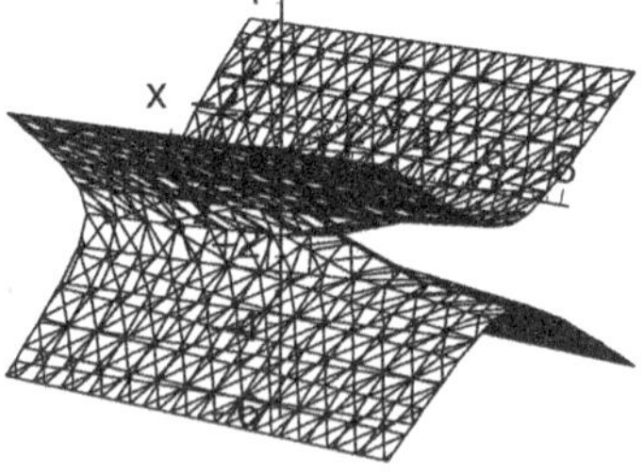

1.3.3 Parametric Curves and Parametric Surfaces

Parametric Curves We have just seen that a line may be parametrized by giving the position (x, y, z) on the line as a function of a parameter t and moreover these functions are linear. More generally, we can parametrize any curve by giving the position (x, y, z) as a function of t not necessarily linear:

$$(x, y, z) = \vec{r}(t) = \big(x(t), y(t), z(t)\big).$$

You can think of t as the time and then $\big(x(t), y(t), z(t)\big)$ is the position of a particle at time t.

Of course, in 2 dimensions, there is no z-component.

Example: 1.16. In $\mathbb{R}^2$, plot the curve parametrized by $\vec{r}(t) = \big(t^2, t^3\big)$ to see it has a "cusp" at $t = 0$. (A cusp is a sharp corner.)

SOLUTION: The curve $\vec{r}(t)$ may be plotted using the `plot` command with a parametric argument:

```
> plot([t^2, t^3, t=-2..2]);
```

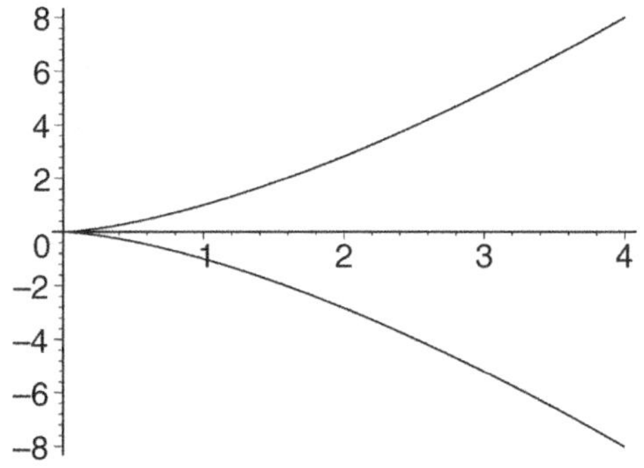

Notice the cusp at the origin.

Example: 1.17. In $\mathbb{R}^3$, plot the helix $\vec{r}(\theta) = \big(6\cos(\theta), 6\sin(\theta), \theta\big)$.

SOLUTION: The helix may be plotted using the `spacecurve` command from the `plots` package:

```
> spacecurve([6*cos(theta), 6*sin(theta), theta], theta=0..6*Pi,
> scaling=constrained, axes=normal, numpoints=73);
```

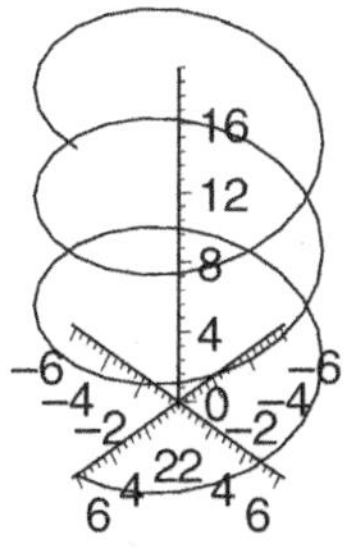

Parametric curves are studied in detail in sections 2.2 and 6.1.

Parametric Surfaces We have also seen that a plane may be parametrized by giving the position (x, y, z) as a linear function of two parameter s and t. We generalize this to a parametrization of any surface by giving the position (x, y, z) as a function of two parameters s and t not necessarily linear:

$$(x, y, z) = \vec{R}(s, t) = \big(x(s, t), y(s, t), z(s, t)\big).$$

Example: 1.18. Plot the parametric surface

$$\vec{R}(\lambda, \theta) = \big(\cosh(\lambda)\cos(\theta), \cosh(\lambda)\sin(\theta), \sinh(\lambda)\big) \ .$$

Then show it is the hyperboloid $x^2 + y^2 - z^2 = 1$.

SOLUTION: We first enter the parametrization into *Maple* as a list of expressions:

```
> R:=[cosh(lambda)*cos(theta), cosh(lambda)*sin(theta),
> sinh(lambda)];
```

$$R := [\cosh(\lambda)\cos(\theta),\ \cosh(\lambda)\sin(\theta),\ \sinh(\lambda)]$$

Then we plot a piece of the surface using the `plot3d` command:

```
> plot3d(R, lambda=-3..2, theta=0..2*Pi);
```

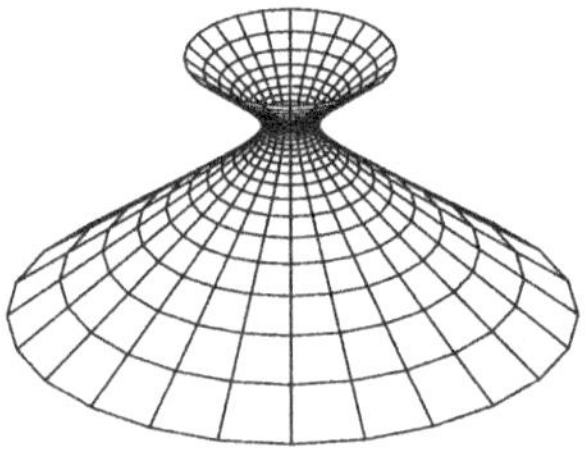

To show it is the hyperboloid, we convert the parametrization into three equations using `equate`:

```
> eqs:=equate([x,y,z], R);
```

$$eqs := \{x = \cosh(\lambda)\cos(\theta),\ y = \cosh(\lambda)\sin(\theta),\ z = \sinh(\lambda)\}$$

and substitute into the equation $x^2 + y^2 - z^2 = 1$ for the hyperboloid:

```
> subs(eqs,x^2 + y^2 - z^2 = 1); simplify(%);
```

$$\cosh(\lambda)^2\cos(\theta)^2 + \cosh(\lambda)^2\sin(\theta)^2 - \sinh(\lambda)^2 = 1$$
$$1 = 1$$

So the equation is satisfied.

Parametric surfaces are studied in detail in Section 6.2.

NOTE: The `plot` command plots **curves** in $\mathbb{R}^2$ either as the graph of a function or in parametric form, while the `implicitplot` command plots **curves** in $\mathbb{R}^2$ in the form of an equation.

Similarly, the `plot3d` command plots **surfaces** in $\mathbb{R}^3$ either as the graph of a function or in parametric form, while the `implicitplot3d` command plots **surfaces** in $\mathbb{R}^3$ in the form of an equation and `spacecurve` plots **curves** in $\mathbb{R}^3$ in parametric form.

1.4 Exercises

- Do Labs: 9.1, 9.2 and 9.3.
- Do Project: 10.1.

1. Consider the vectors

$$\vec{a} = (2,3) \qquad \vec{b} = (-1,2) \qquad \vec{c} = (4,-3)$$
$$\vec{u} = (0,\sqrt{3},1) \qquad \vec{v} = (2,-4,\sqrt{3}) \qquad \vec{w} = (\sqrt{3},1,-2)$$

Compute each of the following quantities:

a) $|\vec{c}|$
b) $|\vec{u}|$
c) $|\vec{v}|$
d) $2\vec{a} - 3\vec{b}$
e) $\sqrt{3}\vec{u} + 2\vec{v}$
f) $\vec{a} \cdot \vec{b}$
g) $\vec{u} \cdot \vec{v}$
h) $\vec{u} \times \vec{v}$
i) the unit vector in the direction of $\vec{c}$
j) the angle between $\vec{a}$ and $\vec{b}$
k) the angle between $\vec{u}$ and $\vec{v}$
l) the vector projection of $\vec{u}$ along $\vec{v}$
m) the projection of $\vec{u}$ orthogonal to $\vec{v}$
n) the area of the triangle with edges $\vec{u}$ and $\vec{v}$
o) the area of the parallelogram with edges $\vec{u}$ and $\vec{v}$
p) the volume of the parallelepiped with edges $\vec{u}$, $\vec{v}$ and $\vec{w}$

2. Repeat problem #1 for the vectors

$$\vec{a} = (1.7,-2.1) \qquad \vec{b} = (-1.4,3.7) \qquad \vec{c} = (4.2,-1.3)$$
$$\vec{u} = (4.1,5.2,3.6) \qquad \vec{v} = (-1.9,2.3,7.2) \qquad \vec{w} = (4.6,-8.3,-6.2)$$

3. In the Earth, Moon and Sun triangle, discussed in example 1.3, find the angle at the Moon when the angle at the Earth is 74.1°. Give the angle in degrees to the nearest hundredth of a degree.

4. Hyperspace, the Final Final Frontier: As our navigator through 4-dimensional hyperspace, your current assignment is to find the angle θ (to the nearest tenth of a degree) at the vertex P of the triangle $\triangle PQR$ with vertices $P = (2,-5,4,-3)$, $Q = (-3,1,0,-2)$ and $R = (5,2,-4,1)$. (The angles are in degrees for the primitive Earthlings.)

5. A 5 kg mass slides 10 m down a frictionless plane which is inclined at a 30° angle from the horizontal. Find the work done on the mass by the force of gravity $\vec{F} = -mg\hat{\jmath}$. Note: $g = 9.8$ M/sec^2.

6. Consider the points $P = (3,4,-2)$, $Q = (0,-3,1)$ and $R = (-2,1,3)$.

   a) Find the parametric equations of the line through P and Q.

   b) Find the symmetric equations of the line through P and Q.

   c) Find the parametric equations of the plane through P, Q and R.

   d) Find the non-parametric equations of the plane through P, Q and R.

7. Consider the points $A = (2,3)$ and $B = (-4,5)$.

   a) Find the parametric equations of the line through A and B.

   b) Find the non-parametric equation of the line through A and B.

8. Find the distance from the point $R = (3, 7)$ to the line $y = 4x - 2$.
Hint: Find two points P and Q on the line. Then the projection of $\overrightarrow{PR}$ orthogonal to $\overrightarrow{PQ}$ is the vector from R to the line which is perpendicular to the line.

9. Find the distance from the point $R = (2, -5, 6)$ to the line $(x, y, z) = (2-t, 4+3t, 1-5t)$.

10. Find the distance from the point $R = (2, -5, 6)$ to the plane $2x + 3y - z = 5$.
Hint: Find the line perpendicular to the plane which passes through R. Then find the foot of this perpendicular.

11. Where does the curve $\vec{r}(t) = (-3t + 5, t^2 - \frac{4}{3}, 2t^2 - \frac{7}{5})$ intersect the yz-plane?

12. Where does the line $\vec{r}(t) = (-3t + 5, t - \frac{4}{3}, 2t - \frac{7}{5})$ intersect the plane $2x + 3y + 4z = 5$?

13. Where does the curve $\vec{r}(t) = (-3t^2+5, t - \frac{4}{3}, 2t^2 - \frac{7}{5})$ intersect the plane $2x+3y+4z = 5$?

14. Plot the parametric curve $\vec{r}(t) = (\cos(5\theta), \sin(3\theta))$ for $0 \le \theta \le 2\pi$. Try changing the 5 and 3 to other integers. What happens? From such a plot, how would you determine the integers? These plots are called Lissajous figures.

15. Plot the parametric curve $x = (2 + \cos 24\theta)\cos\theta$, $y = (2 + \cos 24\theta)\sin\theta$, $z = \sin 24\theta$ for $0 \le \theta \le 2\pi$. To get a good plot, add the option `numpoints=500`.

16. Plot the parametric surface $x = (2 + \cos\phi)\cos\theta$, $y = (2 + \cos\phi)\sin\theta$, $z = \sin\phi$ for $0 \le \phi \le 2\pi$ and $0 \le \theta \le 2\pi$. What shape is the surface?

Chapter 2

Vector Functions of One Variable: Curves

2.1 Vector Functions of One Variable

Remember to restart the `VecCalc` package. See Appendix B.

2.1.1 Definition

A vector-valued function of one variable is an ordered list of real valued functions. In particular, in $\mathbb{R}^3$, a vector valued function has the form

$$\vec{f}(t) = \langle f_1(t), f_2(t), f_3(t) \rangle \ .$$

The independent variable, in this case t, is called the parameter.

Some *Maple* commands expect the function to be entered as an expression, while others expect an arrow-defined function. For example, to enter the vector valued function $\vec{r}(t) = \langle 6\cos(t), 6\sin(t), t \rangle$ as an expression, type:

```
> r:=<6*cos(t)|6*sin(t)|t>;
```

$$r := [6\cos(t),\ 6\sin(t),\ t]$$

To enter it as an arrow-defined function, use the `VecCalc` command `MakeFunction` (or its alias `MF`)

```
> R:=MF(t, <6*cos(t)|6*sin(t)|t>);
```

$$R := [t \to 6\cos(t),\ t \to 6\sin(t),\ t \to t]$$

or the operator `&->` (read "and arrow"):

```
> R:=t &-> <6*cos(t)|6*sin(t)|t>;
```

$$R := [t \to 6\cos(t),\ t \to 6\sin(t),\ t \to t]$$

To evaluate the expression `r` at $t = \dfrac{\pi}{2}$, use the `subs` command:

```
> subs(t=Pi/2,r);
```

$$\left[0,\ 6,\ \frac{\pi}{2}\right]$$

To evaluate the arrow-defined function R at $t = \frac{\pi}{2}$, use the VecCalc command evalFunction (or its alias EF):

```
> EF(R,Pi/2);
```

$$\left[0,\ 6,\ \frac{\pi}{2}\right]$$

or the operator &@ (read "and at"):

```
> R&@(Pi/2);
```

$$\left[0,\ 6,\ \frac{\pi}{2}\right]$$

You can easily convert back and forth between expressions and arrow-defined functions. Given the expression r, the arrow-defined function is

```
> R:=t &-> r;
```

$$R := [t \to 6\cos(t),\ t \to 6\sin(t),\ t \to t]$$

Conversely, given the arrow-defined function R, the expression is

```
> r:=R &@ t;
```

$$r := [6\cos(t),\ 6\sin(t),\ t]$$

Quite often a vector valued function is interpreted as a curve, giving the position as a function of time. For example, the vector valued function $\vec{r}(t)$, defined above, is a helix as was shown in example 1.17. However, a vector valued function can also represent many other physical quantities such as the velocity $\vec{v}(t)$ along a curve as a function of time or the force $\vec{F}(t)$ applied to a particle as a function of time. (See the Figure.)

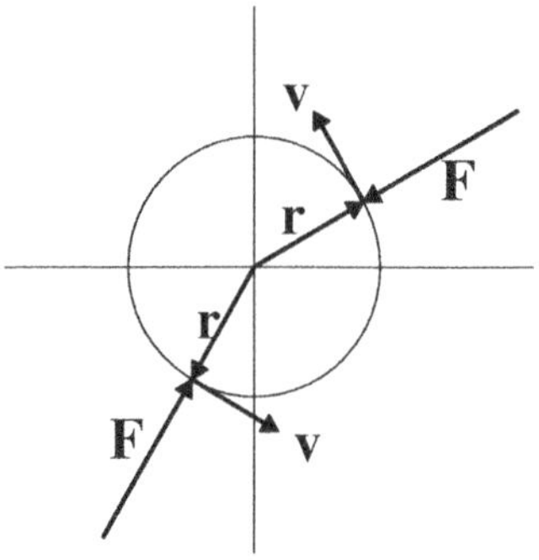

Further, the parameter need not represent time. For example, the standard parametrization of a circle of radius 2 is $\vec{r}(\theta) = \langle 2\cos(\theta), 2\sin(\theta)\rangle$ where the parameter θ measures the angle counter-clockwise from the positive x-axis. After entering the function,

```
> r:=<2*cos(theta)| 2*sin(theta)>;
```

$$r := [2\cos(\theta),\ 2\sin(\theta)]$$

we can plot three quarters of a circle:

```
> plot([r[1], r[2], theta=0..3*Pi/2], scaling=constrained);
```

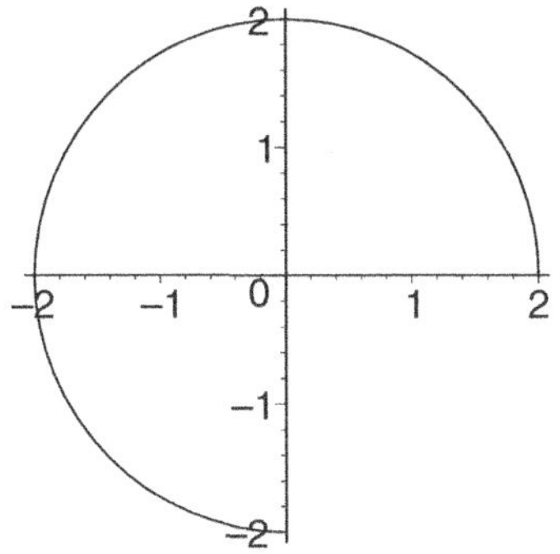

2.1.2 Limits, Derivatives and Integrals

A limit, derivative or integral of a vector-valued function is computed by applying the operation to each component of the vector valued function. For example, the limit as $t \to 2$ of the vector valued function $f(t) = \left\langle \frac{t^2-4}{t-2}, \frac{t^2-5t+6}{t-2} \right\rangle$ is

$$\lim_{t\to 2} \left\langle \frac{t^2-4}{t-2}, \frac{t^2-5t+6}{t-2} \right\rangle = \left\langle \lim_{t\to 2} \frac{t^2-4}{t-2}, \lim_{t\to 2} \frac{t^2-5t+6}{t-2} \right\rangle = \langle 4, -1 \rangle$$

In *Maple* to compute the above limit, we enter the vector valued function as an expression:

```
> f := <(t^2-4)/(t-2)| (t^2-5*t+6)/(t-2)>;
```

$$f := \left[\frac{t^2-4}{t-2}, \frac{t^2-5\,t+6}{t-2} \right]$$

and then use the `Limit` and `value` commands:

```
> Limit(f, t=2);    value(%);
```

$$\left[\lim_{t\to 2} \frac{t^2-4}{t-2}, \lim_{t\to 2} \frac{t^2-5\,t+6}{t-2} \right]$$

$$[4, -1]$$

Notice that the `Limit` command displays the limit while the `value` command computes it. You can also compute this using the `limit` command:

```
> limit(f, t=2);
```

$$[4, -1]$$

but it is better to use `Limit` and `value` because then you get to see the input and check it for typographical errors.

Similarly, the derivative of the curve $\vec{r}(t) = \langle t\cos(t), t\sin(t), t \rangle$ is computed by using the `Diff` and `value` commands:

```
> r:=<t*cos(t)| t*sin(t)| t>;
```

$$r := [t\cos(t),\, t\sin(t),\, t]$$

```
> Diff(r, t);    v_expr:=value(%);
```

$$\left[\frac{d}{dt}\left(t\cos(t)\right),\ \frac{d}{dt}\left(t\sin(t)\right),\ \frac{d}{dt}\,t\right]$$

$$v\_expr := [\cos(t) - t\sin(t),\ \sin(t) + t\cos(t),\ 1]$$

or the `diff` command:

```
> diff(r, t);
```

$$[\cos(t) - t\sin(t),\ \sin(t) + t\cos(t),\ 1]$$

As will be seen in the next section, this vector may be interpreted as the tangent vector to the curve (or its velocity). Its value at $t = \frac{\pi}{2}$ may be obtained from

```
> subs(t=Pi/2,v_expr);
```

$$\left[-\frac{\pi}{2},\ 1,\ 1\right]$$

Alternatively, you can convert the vector of expressions `v_expr` into a vector of arrow defined functions:

```
> v:=t&->v_expr;
```

$$v := [t \to \cos(t) - t\sin(t),\ t \to \sin(t) + t\cos(t),\ t \to 1]$$

and evaluate at $\frac{\pi}{2}$

```
> v&@(Pi/2);
```

$$\left[-\frac{\pi}{2},\ 1,\ 1\right]$$

Turning to integrals, given the vector valued function,

```
> f:=<t| t^2| t^3>;
```

$$f := [t,\ t^2,\ t^3]$$

as an expression, its indefinite integral is computed using the `Int` and `value` commands:

```
> Int(f,t);    value(%);
```

$$\left[\int t\,dt,\ \int t^2\,dt,\ \int t^3\,dt\right]$$

$$\left[\frac{t^2}{2},\ \frac{t^3}{3},\ \frac{t^4}{4}\right]$$

or the `int` command:

```
> int(f,t);
```

$$\left[\frac{t^2}{2},\ \frac{t^3}{3},\ \frac{t^4}{4}\right]$$

The definite integral from $t = 2$ to $t = 3$ is computed similarly:

```
> Int(f,t=2..3);    value(%);
```

$$\left[\int_2^3 t\,dt,\ \int_2^3 t^2\,dt,\ \int_2^3 t^3\,dt\right]$$

$$\left[\frac{5}{2}, \frac{19}{3}, \frac{65}{4}\right]$$

or

```
> int(f,t=2..3);
```

$$\left[\frac{5}{2}, \frac{19}{3}, \frac{65}{4}\right]$$

TIP: Again, it is better to use `Limit` and `value`, `Diff` and `value` or `Int` and `value`, because you can see what you are computing and check for typographical errors.

Notice that the commands `Limit`, `Diff` and `Int` act on an expression. However, when doing derivatives, we often use `D` to differentiate an arrow-defined function. So to differentiate the helix curve

```
> r:=t &-> <t*cos(t)|t*sin(t)|t>;
```

$$r := [t \to t\cos(t),\ t \to t\sin(t),\ t \to t]$$

we simply execute

```
> v:=D(r);
```

$$v := [t \to \cos(t) - t\sin(t),\ t \to \sin(t) + t\cos(t),\ 1]$$

Then the value at $t = \frac{\pi}{2}$ is

```
> v&@(Pi/2);
```

$$\left[-\frac{\pi}{2}, 1, 1\right]$$

and the expression form of this derivative is

```
> v&@t;
```

$$[\cos(t) - t\sin(t),\ \sin(t) + t\cos(t),\ 1]$$

In many ways, this is simpler than using `Diff`.

2.2 Frenet Analysis of Curves

The `VecCalc` package has many commands which simplify the computations in the analysis of a curve. In discussing these quantities, we first cover the details of the computation and then give the `VecCalc` shortcut. You should never use these shortcuts until you fully understand how the quantities are computed. Rather, you should work out the computations and then check with the `VecCalc` command.

In studying the properties of a curve, we will repeatedly refer to two examples, one in $\mathbb{R}^2$ and one in $\mathbb{R}^3$.

- In $\mathbb{R}^2$ we will consider the ellipse $\frac{x^2}{16} + \frac{y^2}{9} = 1$ which may be parametrized by

$$x = 4\cos(\phi) \qquad y = 3\sin(\phi) .$$

It should be noted that the parameter ϕ does not measure angles like the angular coordinate θ of polar coordinates. Nevertheless it does start at zero on the positive x-axis, it does increase as you move counterclockwise around the ellipse and it does increase by $\frac{\pi}{2}$ as you pass through each quadrant.

- In $\mathbb{R}^3$ we will consider the helix parametrized by

$$x = 6\cos(t) \qquad y = 6\sin(t) \qquad z = t\ .$$

Notice that x and y are expressed in terms of polar coordinates on a circle of radius 6 traversed counterclockwise in time and z increases with time.

Curves can also be constructed in higher dimensions, but they are harder to visualize. Further some of the quantities computed below are only defined in $\mathbb{R}^3$ (those which depend on the cross product).

2.2.1 Position and Plot

We input the curves into *Maple* as vectors of expressions. To plot a two dimensional curve, we use the `plot` command with a parametric argument. To plot a three dimensional curve, we use the `spacecurve` command.

Example: 2.1. Plot the ellipse $\vec{r}(\phi) = \langle 4\cos(\phi), 3\sin(\phi)\rangle$.

SOLUTION: For the ellipse, we enter the parametrization

```
> r:=<4*cos(phi)|3*sin(phi)>;
```

$$r := [4\cos(\phi),\ 3\sin(\phi)]$$

which is also called the position vector on the ellipse. To plot the ellipse, we use the `plot` command:

```
> plot([r[1], r[2], phi=0..2*Pi], scaling=constrained);
```

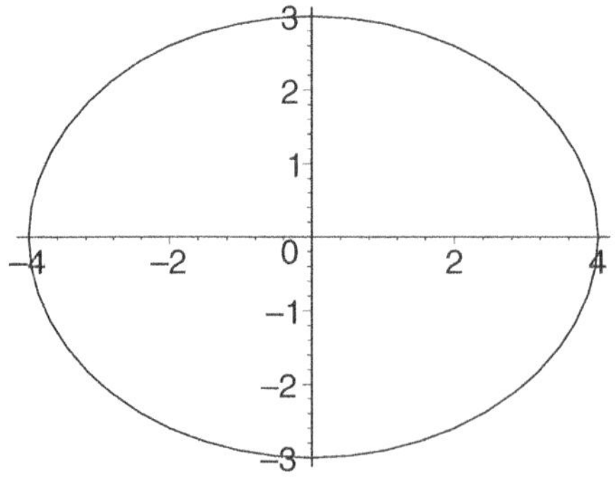

NOTE: The option `scaling=constrained` makes sure the same scale is used on both axes.

Example: 2.2. Plot the helix $\vec{R}(t) = (6\cos(t), 6\sin(t), t)$.

SOLUTION: For the helix, we enter the curve (or position):

```
> R:=<6*cos(t)|6*sin(t)|t>;
```

$$R := [6\cos(t),\, 6\sin(t),\, t]$$

We plot the helix by using the `spacecurve` command:

```
> spacecurve([R[1],R[2],R[3]], t=0..6*Pi,  scaling=constrained,
> axes=normal, numpoints=73);
```

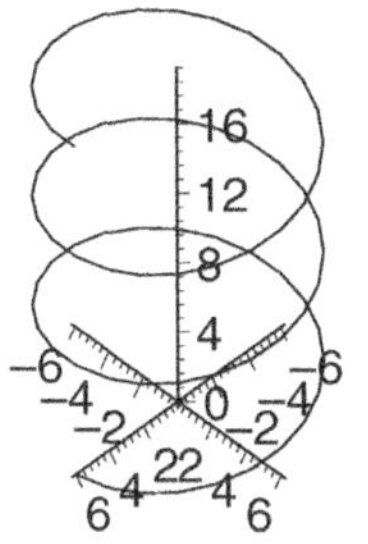

2.2.2 Velocity, Acceleration and Jerk

When the parameter along a curve is interpreted as the time, the derivative of the position is the velocity, the derivative of the velocity is the acceleration and the derivative of the acceleration is the jerk. Even when the parameter is not the time, the words velocity, acceleration and jerk may still be used for the first, second and third derivatives of the position.

If the position has been defined as a vector of expressions, then the derivatives may be computed using `Diff` and `value` or `diff`. The `VecCalc` package also has the commands `CurveVelocity`, `CurveAcceleration` and `CurveJerk` to compute these directly from the position without needing to compute them in order. The aliases are `Cv`, `Ca` and `Cj`.

Example: 2.3. Find the velocity, acceleration and jerk of the ellipse of example 2.1.

SOLUTION: The position was entered in example 2.1. So the velocity, acceleration and jerk are:

```
> v:=Diff(r,phi); v:=value(%);
```

$$v := \left[\frac{d}{d\phi}(4\cos(\phi)),\, \frac{d}{d\phi}(3\sin(\phi))\right]$$

$$v := [-4\sin(\phi),\, 3\cos(\phi)]$$

```
> a:=diff(v,phi); j:=diff(a,phi);
```

$$a := [-4\cos(\phi),\, -3\sin(\phi)]$$

$$j := [4\sin(\phi),\, -3\cos(\phi)]$$

These may be checked using `Cv`, `Ca` and `Cj`:

```
> v:=Cv(r,phi); a:=Ca(r,phi); j:=Cj(r,phi);
```

$$v := [-4\sin(\phi),\, 3\cos(\phi)]$$

$$a := [-4\cos(\phi),\ -3\sin(\phi)]$$
$$j := [4\sin(\phi),\ -3\cos(\phi)]$$

Example: 2.4. Find the velocity, acceleration and jerk of the helix of example 2.2.

SOLUTION: The position was entered in example 2.2. So the velocity, acceleration and jerk are

```
> V:=diff(R,t); A:=diff(V,t); J:=diff(A,t);
```

$$V := [-6\sin(t),\ 6\cos(t),\ 1]$$
$$A := [-6\cos(t),\ -6\sin(t),\ 0]$$
$$J := [6\sin(t),\ -6\cos(t),\ 0]$$

(Notice that we are using the capital letters `R`, `V`, `A` and `J` for the helix, solely to distinguish these quantities from the corresponding quantities for the ellipse `r`.)

2.2.3 Speed, Arc Length and Arc Length Parameter

The length of the velocity is called the speed and may be computed using the `Length` command. The definite integral of the speed is the arc length and the arc length with a variable final point defines the arc length parameter s at the final point. Sometimes the arc length parameter is used to reparametrize the curve.

The `VecCalc` package has a command `CurveSpeed` (or `Cs`) which will compute the speed, and a command `CurveLength` (or `CL`) which will compute the arc length integral as a function of the two endpoints. You can then plug in the endpoints and compute the `value`.

Example: 2.5. For the helix of example 2.2, find the speed and arc length around one cycle. Then find the arc length parameter and reparametrize the helix in terms of the arc length parameter, if possible.

SOLUTION: Using the velocity computed in the previous example, the speed is

```
> SPEED:=Length(V);
```

$$SPEED := \sqrt{37}$$

and the arc length around one cycle is

```
> Int(SPEED, t=0..2*Pi); value(%);
```

$$\int_0^{2\pi} \sqrt{37}\, dt$$
$$2\sqrt{37}\,\pi$$

The arc length parameter s is

```
> Int(SPEED, t=0..T); s_param:=value(%);
```

$$\int_0^{T} \sqrt{37}\, dt$$

$$s\_param := \sqrt{37}\, T$$

where T is the time at the final point. We can reparametrize the curve in terms of an arc length parameter s by solving the equation `s = s_param` for T and plugging into the curve $\vec{R}(t)$:

```
> Tsol:=solve(s=s_param,T);
```

$$Tsol := \frac{s\sqrt{37}}{37}$$

```
> R;
```

$$[6\cos(t),\, 6\sin(t),\, t]$$

```
> subs(t=Tsol,R);
```

$$\left[6\cos(\frac{s\sqrt{37}}{37}),\, 6\sin(\frac{s\sqrt{37}}{37}),\, \frac{s\sqrt{37}}{37}\right]$$

We can check some of these results using `Cs` and `CL`:

```
> Cs(R,t);
```

$$\sqrt{37}$$

```
> L:=CL(R,t); L(0,2*Pi); value(%);
```

$$L := (a, b) \to \int_a^b \sqrt{37}\, dt$$

$$\int_0^{2\pi} \sqrt{37}\, dt$$

$$2\sqrt{37}\,\pi$$

```
> L(0,T); s=value(%);
```

$$\int_0^{T} \sqrt{37}\, dt$$

$$s = \sqrt{37}\, T$$

Example: 2.6. For the ellipse of example 2.1, find the speed and arc length once around. Then find the arc length parameter and reparametrize the ellipse in terms of the arc length parameter, if possible.

SOLUTION: Using the velocity computed in a previous example, we compute the speed and the arc length once around:

```
> speed:=Length(v);
```

$$speed := \sqrt{-7\cos(\phi)^2 + 16}$$

```
> Int(speed, phi=0..2*Pi);   value(%);
```

$$\int_0^{2\pi} \sqrt{-7\cos(\phi)^2 + 16}\, d\phi$$

$$16\,\text{EllipticE}(\frac{\sqrt{7}}{4})$$

Notice that the `value` command gave the answer in terms of the elliptic E function. This is not very informative. So we use the `evalf` command to get a numerical approximation:

```
> evalf(%);
```

$$22.10349216$$

Using the `VecCalc` commands `Cs` and `CL`, we check:

```
> Cs(r,phi);
```

$$\sqrt{-7\cos(\phi)^2+16}$$

```
> L:=CL(r,phi);    L(0,2*Pi);
```

$$L := (a, b) \rightarrow \int_a^b \sqrt{-7\cos(\phi)^2+16}\,d\phi$$

$$\int_0^{2\pi} \sqrt{-7\cos(\phi)^2+16}\,d\phi$$

To find the arc length parameter, we need to compute the integral

```
> L(0,T);    s := value(%);
```

$$\int_0^{T} \sqrt{-7\cos(\phi)^2+16}\,d\phi$$

$$s := \int_0^{T} \sqrt{-7\cos(\phi)^2+16}\,d\phi$$

Maple is unable to compute this integral. However, if necessary, we can always work with it implicitly or numerically, since we know the derivative:

```
> diff(s,T);
```

$$\sqrt{-7\cos(T)^2+16}$$

2.2.4 Unit Tangent, Unit Principal Normal, Unit Binormal

The unit tangent vector $\hat{T}$ along a curve $\vec{r}(t)$ is the unit vector in the direction of the velocity $\vec{v}$:

$$\hat{T} = \hat{v} = \frac{\vec{v}}{|\vec{v}|}.$$

The `VecCalc` command is `CurveTangent` (or `CT`).

The unit (principal) normal vector $\hat{N}$ along a curve $\vec{r}(t)$ is the unit vector perpendicular to the velocity $\vec{v}$ in the plane of the velocity and the acceleration $\vec{a}$ on the same side of the velocity as the acceleration. You compute it by finding the projection of $\vec{a}$ perpendicular to $\vec{v}$ or $\hat{T}$ and then dividing by its length:

$$\text{proj}_{\perp\vec{v}}\vec{a} = \vec{a} - \text{proj}_{\vec{v}}\vec{a} = \vec{a} - (\vec{a}\cdot\hat{T})\hat{T}$$

$$\hat{N} = \frac{\text{proj}_{\perp\vec{v}}\vec{a}}{|\text{proj}_{\perp\vec{v}}\vec{a}|}\ .$$

The `VecCalc` command is `CurveNormal` (or `CN`).

In $\mathbb{R}^3$, you can also compute the unit binormal vector $\hat{B}$ which is a unit vector perpendicular to $\hat{T}$ and $\hat{N}$ and given by $\hat{B} = \hat{T} \times \hat{N}$. Equivalently, $\hat{B}$ is the unit vector perpendicular to $\vec{v}$ and $\vec{a}$ and given by

$$\hat{B} = \frac{\vec{v} \times \vec{a}}{|\vec{v} \times \vec{a}|} .$$

This latter formula is the easiest way to compute $\hat{B}$ because you don't need to first compute $\hat{N}$. Further, $\hat{N}$ can then be computed in $\mathbb{R}^3$ from the formula

$$\hat{N} = \hat{B} \times \hat{T}$$

which is obtained from $\hat{B} = \hat{T} \times \hat{N}$ by cyclically permuting the three unit vectors. The `VecCalc` command is `CurveBinormal` (or `CB`).

Example: 2.7. Find the unit tangent and unit normal vectors of the ellipse of example 2.1.

NOTE: Since the ellipse is 2-dimensional, there is no binormal.

SOLUTION: Using the velocity and acceleration computed in previous examples, we have

```
> t_hat:=simplify(v/speed);
```

$$t\_hat := \left[-\frac{4\sin(\phi)}{\sqrt{-7\cos(\phi)^2+16}}, \frac{3\cos(\phi)}{\sqrt{-7\cos(\phi)^2+16}}\right]$$

```
> perp_proj:=(a-Dot(a,t_hat)*t_hat):  perp_proj:=simplify(%);
```

$$perp\_proj := \left[\frac{36\cos(\phi)}{7\cos(\phi)^2-16}, \frac{48\sin(\phi)}{7\cos(\phi)^2-16}\right]$$

```
> perp_proj/Length(perp_proj):  n_hat:=simplify(%);
```

$$n\_hat := \left[-\frac{3\cos(\phi)}{\sqrt{-7\cos(\phi)^2+16}}, -\frac{4\sin(\phi)}{\sqrt{-7\cos(\phi)^2+16}}\right]$$

and we check with the `VecCalc` commands:

```
> CT(r,phi); CN(r,phi);
```

$$\left[-\frac{4\sin(\phi)}{\sqrt{-7\cos(\phi)^2+16}}, \frac{3\cos(\phi)}{\sqrt{-7\cos(\phi)^2+16}}\right]$$

$$\left[-\frac{3\cos(\phi)}{\sqrt{-7\cos(\phi)^2+16}}, -\frac{4\sin(\phi)}{\sqrt{-7\cos(\phi)^2+16}}\right]$$

Example: 2.8. Find the unit tangent, unit normal and unit binormal vectors of the helix of example 2.2.

SOLUTION: Using the velocity and acceleration computed in previous examples, we have

```
> T:=V/Length(V);
```

$$T := \left[-\frac{6}{37}\sqrt{37}\sin(t), \frac{6}{37}\sqrt{37}\cos(t), \frac{\sqrt{37}}{37}\right]$$

```
> VxA:=Cross(V,A);
```

$$VxA := [6\sin(t),\ -6\cos(t),\ 36]$$

```
> B:=VxA/Length(VxA);
```

$$B := \left[\frac{1}{37}\sqrt{37}\sin(t),\ -\frac{1}{37}\sqrt{37}\cos(t),\ \frac{6\sqrt{37}}{37}\right]$$

```
> N:=B &x T;
```

$$N := [-\cos(t),\ -\sin(t),\ 0]$$

and we check with the VecCalc commands:

```
> CT(R,t); CB(R,t); CN(R,t);
```

$$\left[-\frac{6}{37}\sqrt{37}\sin(t),\ \frac{6}{37}\sqrt{37}\cos(t),\ \frac{\sqrt{37}}{37}\right]$$

$$\left[\frac{1}{37}\sqrt{37}\sin(t),\ -\frac{1}{37}\sqrt{37}\cos(t),\ \frac{6\sqrt{37}}{37}\right]$$

$$[-\cos(t),\ -\sin(t),\ 0]$$

In $\mathbb{R}^4$ and higher dimensions, there are generalizations of $\hat{T}$, $\hat{N}$ and $\hat{B}$ but they cannot be computed using the cross product. Rather they are computed from the velocity and successively higher derivatives of the curve by applying the Gramm-Schmidt procedure. But that is a topic for a course in linear algebra.

2.2.5 Curvature and Torsion

The curvature κ along a curve $\vec{r}(t)$ measures the rate at which the direction of the curve is changing and may be computed from any of the formulas:

$$\kappa = \left|\frac{d\hat{T}}{ds}\right| = \frac{1}{|\vec{v}|}\left|\frac{d\hat{T}}{dt}\right| = \frac{\vec{a}\cdot\hat{N}}{|\vec{v}|^2} = \frac{|\vec{v}\times\vec{a}|}{|\vec{v}|^3}\ .$$

Since the last formula involves a cross product, it can only be used in $\mathbb{R}^3$. The `VecCalc` command is `CurveCurvature` (or `Ck`).

The torsion τ along a curve $\vec{r}(t)$ measures the rate at which the plane of the curve is changing and may be computed from either of the formulas:

$$\tau = -\frac{d\hat{B}}{ds}\cdot\hat{N} = \frac{(\vec{v}\times\vec{a})\cdot\vec{j}}{|\vec{v}\times\vec{a}|^2}\ .$$

Since the definition involves $\hat{B}$, the torsion is only defined in $\mathbb{R}^3$. The `VecCalc` command is `CurveTorsion` (or `Ct`).

Example: 2.9. Find the curvature of the ellipse of example 2.1.

NOTE: Since the ellipse is 2-dimensional, there is no torsion.

SOLUTION: Using the acceleration and unit normal from previous examples, we compute

```
> kappa:=Dot(a,n_hat)/speed^2;
```

$$\kappa := \frac{12}{(-7\cos(\phi)^2 + 16)^{(3/2)}}$$

and we check with the VecCalc commands:

```
> Ck(r,phi);
```

$$\frac{12}{(-7\cos(\phi)^2 + 16)^{(3/2)}}$$

Example: 2.10. Find the curvature and torsion of the helix of example 2.2.

SOLUTION: Using quantities from previous examples, we compute

```
> Kappa:=Length(VxA)/SPEED^3;
```

$$K := \frac{6}{37}$$

```
> Tau:=Dot(VxA,J)/Length(VxA)^2;
```

$$T := \frac{1}{37}$$

and we check with the VecCalc commands:

```
> Ck(R,t); Ct(R,t);
```

$$\frac{6}{37}$$
$$\frac{1}{37}$$

2.2.6 Tangential and Normal Components of Acceleration

Since the acceleration $\vec{a}$ lies in the plane of the vectors $\hat{T}$ and $\hat{N}$, we can write it as

$$\vec{a} = a_T\hat{T} + a_N\hat{N} .$$

Since $\hat{T}$ and $\hat{N}$ are perpendicular unit vectors, we can identify the coefficients, a_T and a_N, as the components of $\vec{a}$ along $\hat{T}$ and $\hat{N}$. These are also called the tangential and normal accelerations. They may be computed from the formulas

$$a_T = \vec{a}\cdot\hat{T} = \frac{d\,|\vec{v}|}{dt} \quad \text{and} \quad a_N = \vec{a}\cdot\hat{N} = \kappa\,|\vec{v}|^2$$

The VecCalc commands are CurveTangentialAcceleration (or CaT) and CurveNormalAcceleration (or CaN).

Example: 2.11. Find the tangential and normal accelerations for the ellipse of example 2.1.

SOLUTION: Using the speed and curvature of the ellipse found in previous examples, we compute

```
> a_T:=diff(speed,phi);
```

$$a\_T := \frac{7\cos(\phi)\sin(\phi)}{\sqrt{-7\cos(\phi)^2+16}}$$

```
> a_N:=kappa*speed^2;
```

$$a\_N := \frac{12}{\sqrt{-7\cos(\phi)^2+16}}$$

and we check with the VecCalc commands:

```
> CaT(r,phi);   CaN(r,phi);
```

$$\frac{7\cos(\phi)\sin(\phi)}{\sqrt{-7\cos(\phi)^2+16}}$$

$$\frac{12}{\sqrt{-7\cos(\phi)^2+16}}$$

Example: 2.12. Find the tangential and normal accelerations for the helix of example 2.2.

SOLUTION: Using the acceleration, unit tangent and unit normal vectors of the helix found in previous examples, we compute (using a different method)

```
> A_T:=A&.T;
```

$$A\_T := 0$$

```
> A_N:=A&.N;
```

$$A\_N := 6$$

and we check with the VecCalc commands:

```
> CaT(R,t);   CaN(R,t);
```

$$0$$
$$6$$

2.3 Exercises

- Do Labs: 9.4 and 9.5.
- Do Project: 10.2.

NOTE: You should only use the VecCalc Curve commands (Cv, Ca, Cj, Cs, CL, CT, CN, CB, Ck, Ct, CaT and CaN) to check your work, except as indicated. Be sure to simplify your answers.

1. Consider the 3-dimensional parametrized curve $\vec{r}(t) = \left(t\cos(t), t\sin(t), \frac{t^3}{6}\right)$:

   (a) Enter the curve into *Maple* as a vector of expressions.

   (b) Plot the curve for $0 \le t \le 2\pi$ with normal axes.

(c) For general times, compute the velocity, acceleration, jerk, speed, unit tangent vector, unit binormal vector, unit normal vector, curvature, torsion, tangential acceleration (in two ways) and normal acceleration (in two ways). Check with the `VecCalc Curve` commands.

(d) Compute the length of the curve for $0 \le t \le 2\pi$.

(e) Find the time when the curvature is a maximum. What is the maximum curvature?

2. The Astroid: Consider the 2-dimensional parametrized curve $\vec{r}(t) = \left(\cos^3(t), \sin^3(t)\right)$:

(a) Assume t is real by executing

```
> assume(t::real);
```

Then enter the curve into *Maple* as a vector of expressions.

NOTE: See Appendix A for information on eliminating the tildes (˜) following the `t`'s.

(b) Plot the curve for $0 \le t \le 2\pi$ to see that it is star shaped. This is why it is called an astroid.

(c) For general times, compute the velocity, acceleration, jerk, speed, unit tangent vector, unit normal vector (Hint: first find the projection of $\vec{a}$ orthogonal to $\vec{v}$.), curvature, tangential acceleration and normal acceleration. Check with the `VecCalc Curve` commands. If they don't look the same, `simplify` the difference between the answers to get 0.

(d) Compute the length of the astroid for $0 \le t \le 2\pi$.

3. Spiral Curve: Consider the 2-dimensional parametrized curve $\vec{r}(t) = \left(t\cos(t), t\sin(t)\right)$:

(a) Enter the curve into *Maple* as a vector of expressions.

(b) Plot the curve for $0 \le t \le 6\pi$ to see that it is a spiral.

(c) For general times, compute the velocity. Label it `v`.

(d) Find the first two points ($t > 0$) where the curve is horizontal.

HINT: The curve is horizontal when $v_2 = \dfrac{dy}{dt} = 0$. Use `fsolve` with ranges to find the values of t and `subs` to find the points. For example, to find the first horizontal point, plot `v[2]`, note there is a zero between 1 and 3 and execute

```
> t1:=fsolve(v[2]=0,t=1..3);
> subs(t=t1,r);
```

(e) Find the first two points where the curve is vertical.

HINT: The curve is vertical when $v_1 = \dfrac{dx}{dt} = 0$.

(f) Compute the length of the spiral for $0 \le t \le 2\pi$ and for $2\pi \le t \le 4\pi$. You may use `CL`. Give both exact and decimal answers.

(g) Plot the curve for $-\dfrac{3\pi}{2} \le t \le \dfrac{3\pi}{2}$ with the options `filled=true, axes=none, color=red` to make a Valentine's card. Right click in the plot to save it in a form for mailing or posting on your website.

4. Find parametric equations for the line tangent to the curve $\vec{r}(t) = (\cos(3t), \sin(-5t))$ at the point $r\left(\dfrac{\pi}{4}\right) = \left(\dfrac{-1}{\sqrt{2}}, \dfrac{1}{\sqrt{2}}\right)$. Plot the curve and the tangent line using `plot` and combine them into one plot using `display`.

5. Find the tangent line to the curve $\vec{r}(t) = (\cos(-6t), 4t, \sin(2t))$ at the point $(0, \pi, 1)$. Plot the curve and the tangent line using `spacecurve` and combine them into one plot using `display`. Rotate the plot with your mouse to get a good view.

6. Find the limiting velocity as $t \to \infty$ on the curve $r(t) = \left(4t + 3\ln\left(\frac{2t}{t+1}\right), -3t + \ln\left(\frac{2t}{t+1}\right)\right)$.

7. The electric force on a point charge q due to a point charge Q is $\vec{F} = -\frac{kqQ}{r^3}\vec{r}$ where $\vec{r}$ is the vector from q to Q and $r = |\vec{r}|$. A small piece of a wire of length $ds = |\vec{v}|dt$ and linear charge density ρ may be approximated as a point charge $dQ = \rho ds$. Calculate the electric force on a point charge of q coulombs located at the origin due to a constant linear charge density ρ along the circle $\vec{r}(t) = (3\cos(t), 3\sin(t), 4)$ for $0 \le t \le 2\pi$.
HINT: The total force is the sum (integral) of the forces due to each of the small pieces of wire.

Chapter 3

Partial Derivatives

3.1 Scalar Functions of Several Variables

3.1.1 Definition

A scalar-valued function of several variables is a real valued function of n variables. Typical examples might be the temperature on a metal plate or the density of a solid. For example, the temperature of the air near a candle might be given by the scalar function of 3 variables, $T(x, y, z) = 300e^{-x^2-y^2-z^2}$. You may enter this function into *Maple* either as an expression

```
> T:=300*exp(-x^2-y^2-z^2);
```

$$T := 300\, e^{(-x^2-y^2-z^2)}$$

or as an arrow-defined function either explicitly by typing:

```
> T:=(x,y,z) -> 300*exp(-x^2-y^2-z^2);
```

$$T := (x,\, y,\, z) \to 300\, e^{(-x^2-y^2-z^2)}$$

or by using the `VecCalc` command `MakeFunction` or its alias `MF`:

```
> T:=MF(<x,y,z>, 300*exp(-x^2-y^2-z^2));
```

$$T := (x,\, y,\, z) \to 300\, e^{(-x^2-y^2-z^2)}$$

or the equlvalent operator `&->` :

```
> T:=<x,y,z> &-> (300*exp(-x^2-y^2-z^2));
```

$$T := (x,\, y,\, z) \to 300\, e^{(-x^2-y^2-z^2)}$$

(Don't forget the extra parentheses following `&->` ; else you may get strange results.)

What is the difference between these two arrow definitions? None as defined above. However, suppose you have already defined an expression r which gives the distance from the origin:

```
> r:=sqrt(x^2+y^2+z^2);
```

$$r := \sqrt{x^2 + y^2 + z^2}$$

and you want to define T in terms of r. If you use the explicit arrow definition,

```
> T:=(x,y,z) -> 300*exp(-r^2);
```

$$T := (x,\, y,\, z) \to 300\, e^{(-r^2)}$$

then *Maple* does not evaluate `r`. Worse still, *Maple* may give the *wrong answer* if you evaluate `T`:

```
> T(x,y,z), T(1,2,3);
```

$$300\, e^{(-x^2-y^2-z^2)},\ 300\, e^{(-x^2-y^2-z^2)}$$

However, if you use `MF` or `&->` , then *Maple* evaluates `r` immediately:

```
> T:=<x,y,z> &-> (300*exp(-r^2));
```

$$T := (x,\, y,\, z) \to 300\, e^{(-x^2-y^2-z^2)}$$

and evaluates `T` correctly:

```
> T(x,y,z), T(1,2,3);
```

$$300\, e^{(-x^2-y^2-z^2)},\ 300\, e^{(-14)}$$

NOTE: In this context of a scalar function, the `unapply` command gives the same result. However, as we will see later, in the context of a vector function, `MF` and `unapply` commands give differnet results.

3.1.2 Plots

To visualize a function of 2 variables, you can plot its graph using the `plot3d` command. Alternatively, you can look at its level curves using either of two commands, `contourplot` or `contourplot3d`, from the `plots` package (which is autoloaded by `VecCalc`).

The command `contourplot3d` is faster than `contourplot` since it uses the machine's floating point processing. It is entirely equivalent to the `plot3d` command with the option `style=contour` in that it produces the 3 dimensional graph of the function but draws the contour lines on the surface. To see the plot from directly above, you should add the option `orientation=[-90,0]` which gives the location of your eye using the spherical coordinates (θ, ϕ). Thus $[-90, 0]$ means that you are looking down on the plot from the positive z-axis with the x- and y-axes in their usual positions.

The problem with `contourplot3d` is that you cannot superimpose an ordinary 2 dimensional plot onto the contour plot. On the other hand `contourplot` is slower but it produces a true 2 dimensional plot of the contour lines. You can then use the `display` command from the `plots` package to superimpose the graph of a function produced using `plot` or a parametric curve produced using `plot` or the graph of an equation produced using `implicitplot`.

NOTE: These plot commands expect the function to be an expression. If `f` is defined using an arrow definition, then `f(x,y)` is the expression.

Example: 3.1. Enter the functions

$$f(x, y) = \sqrt{(x-2)^2 + y^2} + \sqrt{(x+2)^2 + y^2}$$

and

$$g(x,y) = \sqrt{(x-2)^2 + y^2} - \sqrt{(x+2)^2 + y^2}$$

as expressions. Plot their graphs and contour plots. Then discuss the shape of the contours and the local maxima and minima of the functions. Notice that f is the sum of the distances from (x,y) to the points $(2,0)$ and $(-2,0)$, while g is the difference.

SOLUTION: We enter the function f and draw its graph and its contour plot:

```
> f:=sqrt((x-2)^2 + y^2) + sqrt((x+2)^2 + y^2);
```

$$f := \sqrt{x^2 - 4\,x + 4 + y^2} + \sqrt{x^2 + 4\,x + 4 + y^2}$$

```
> plot3d(f, x=-4..4, y = -3..3, axes=framed, orientation=[60,75]);
```

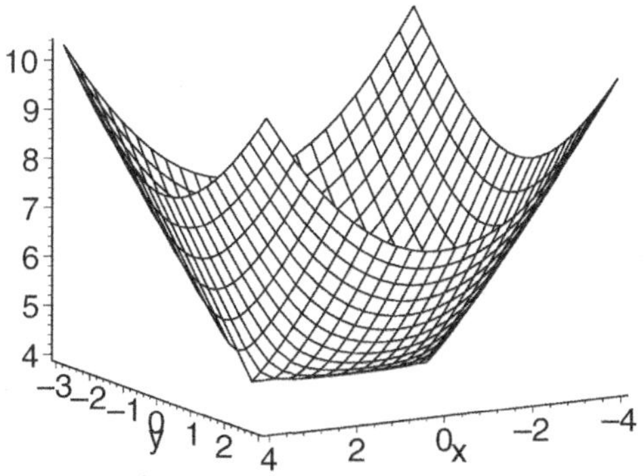


```
> contourplot3d(f, x=-4..4, y = -3..3, axes=framed,
> scaling=constrained, orientation=[-90,0]);
```

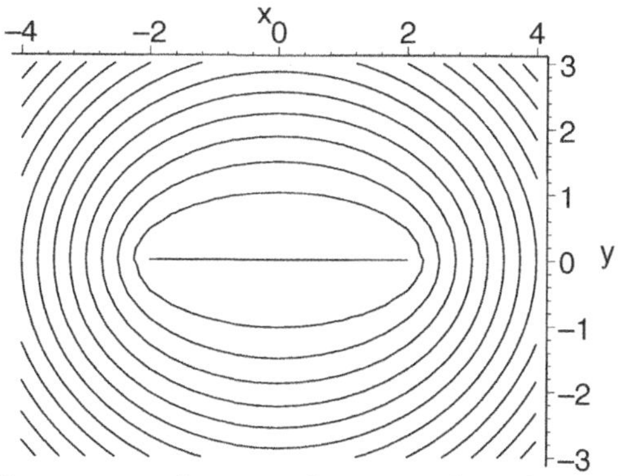

Notice that the line segment between the points $(-2,0)$ and $(2,0)$ is the level set of f with value 4 and the other level sets are ellipses with foci at $(-2,0)$ and $(2,0)$. (This is the definition of an ellipse.) Thus the minimum occurs along the line segment between $(-2,0)$ and $(2,0)$.

Next we enter the function g and draw its graph and its contour plot:

```
> g:=sqrt((x-2)^2 + y^2) - sqrt((x+2)^2 + y^2);
```

$$g := \sqrt{x^2 - 4\,x + 4 + y^2} - \sqrt{x^2 + 4\,x + 4 + y^2}$$

```
> plot3d(g, x=-5..5, y = -5..5, axes=boxed, orientation=[60,75]);
```

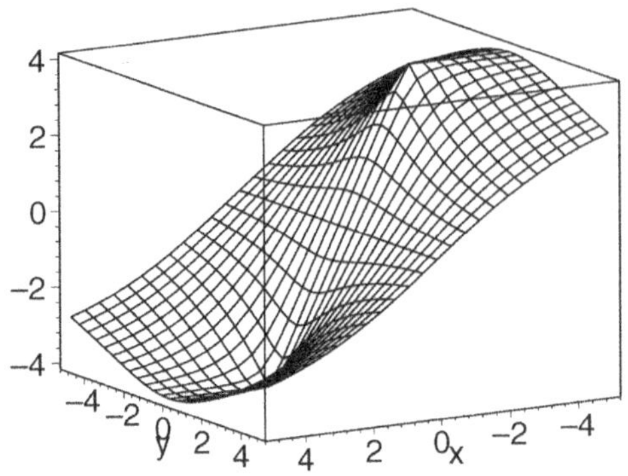


```
> contourplot3d( g, x=-5..5, y = -5..5, axes=boxed,
> orientation=[-90,0]);
```

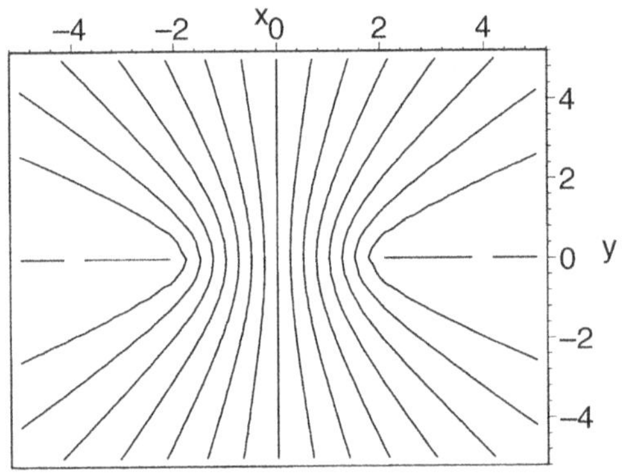

This time the y-axis is the level set of g with value 0, the part of the x-axis with $x > 2$ is the level set with value -4 and the part of the x-axis with $x < -2$ is the level set with value 4. The remaining level sets are half-hyperbolas with foci at $(2, 0)$ and $(-2, 0)$. (This is the definition of a hyperbola.) Thus the minimum occurs along the part of the x-axis with $x > 2$ and the maximum occurs along the part of the x-axis with $x < -2$.

3.1.3 Partial Derivatives

Given a function of several variables, defined as an expression, its partial derivatives may be computed using `Diff` and `value` or simply `diff`. For example, if

```
> f:=cos(x*y+t^2);
```

$$f := \cos(x\,y + t^2)$$

then $\dfrac{\partial f}{\partial y}$ would be computed using

```
> Diff(f,y); value(%);
```

$$\frac{\partial}{\partial y}\cos(x\,y+t^2)$$
$$-\sin(x\,y+t^2)\,x$$

or

```
> diff(f,y);
```

$$-\sin(x\,y+t^2)\,x$$

Higher derivatives may also be computed using additional arguments. For example, $\dfrac{\partial^3 f}{\partial y^2 \partial t}$ would be computed using

```
> diff(f,y,y,t);
```

$$2\sin(x\,y+t^2)\,t\,x^2$$

The results from these commands are expressions.

Given a function of several variables, defined in arrow notation, its partial derivatives may be computed using `D` with an index which is the number of the variable. For example, if

```
> f:=(x,y,t) -> cos(x*y+t^2);
```

$$f := (x,\, y,\, t) \to \cos(y\,x+t^2)$$

then $\dfrac{\partial f}{\partial y}$ would be computed using

```
> D[2](f);
```

$$(x,\, y,\, t) \to -\sin(y\,x+t^2)\,x$$

and $\dfrac{\partial^3 f}{\partial y^2 \partial t}$ would be computed using

```
> D[2,2,3](f);
```

$$(x,\, y,\, t) \to 2\sin(y\,x+t^2)\,x^2\,t$$

The results from these commands are arrow-defined functions.

Example: 3.2. Use arrow notation to enter the function

$$f(x,y,\theta,\phi) = y\sin(\phi)\cos(\theta) - xy\cos(\phi)$$

compute the derivatives $\dfrac{\partial f}{\partial \theta}$ and $\dfrac{\partial^4 f}{\partial y \partial^2 \theta \partial \phi}$ and evaluate them at a point (a,b,t,p).

SOLUTION: We enter the function and use `D` to compute the derivatives:

```
> f:=(x,y,theta,phi) -> y*sin(phi)*cos(theta) - x*y*cos(phi);
```

$$f := (x,\, y,\, \theta,\, \phi) \to y\sin(\phi)\cos(\theta) - x\,y\cos(\phi)$$

```
> f_theta:=D[3](f);
```

$$f\_theta := (x,\, y,\, \theta,\, \phi) \to -y\sin(\phi)\sin(\theta)$$

```
> f_yphiphitheta:=D[2,4,4,3](f);
```

$$f\_yphiphitheta := (x,\, y,\, \theta,\, \phi) \to \sin(\phi)\sin(\theta)$$

Finally, we evaluate at (a, b, t, p).

```
> f(a,b,t,p);
```

$$b\sin(p)\cos(t) - a\,b\cos(p)$$

```
> f_theta(a,b,t,p);
```

$$-b\sin(p)\sin(t)$$

```
> f_yphiphitheta(a,b,t,p);
```

$$\sin(p)\sin(t)$$

Example: 3.3. Enter the functions

$$f(x, y) = \sqrt{(x-2)^2 + y^2} + \sqrt{(x+2)^2 + y^2}$$

and

$$g(x, y) = \sqrt{(x-2)^2 + y^2} - \sqrt{(x+2)^2 + y^2}$$

as expressions. Then compute the quantities $C = \dfrac{\partial g}{\partial x} - \dfrac{\partial f}{\partial y}$ and $L = \dfrac{\partial^2 g}{\partial x^2} + \dfrac{\partial^2 g}{\partial y^2}$ evaluated at a point (a, b).

SOLUTION: The functions are

```
> f:=sqrt((x-2)^2 + y^2) + sqrt((x+2)^2 + y^2):
> g:=sqrt((x-2)^2 + y^2) - sqrt((x+2)^2 + y^2):
```

We then compute the required quantities and evaluate at (a, b):

```
> C:=diff(g,x) - diff(f,y); Cab:=subs(x=a,y=b,C);
```

$$C := \frac{2x-4}{2\sqrt{x^2-4x+4+y^2}} - \frac{2x+4}{2\sqrt{x^2+4x+4+y^2}} - \frac{y}{\sqrt{x^2-4x+4+y^2}} - \frac{y}{\sqrt{x^2+4x+4+y^2}}$$

$$Cab := \frac{2a-4}{2\sqrt{a^2-4a+4+b^2}} - \frac{2a+4}{2\sqrt{a^2+4a+4+b^2}} - \frac{b}{\sqrt{a^2-4a+4+b^2}} - \frac{b}{\sqrt{a^2+4a+4+b^2}}$$

```
> L:=diff(g,x,x) + diff(g,y,y); Lab:=subs(x=a,y=b,L);;
```

$$L := -\frac{(2x-4)^2}{4(x^2-4x+4+y^2)^{(3/2)}} + \frac{2}{\sqrt{x^2-4x+4+y^2}} + \frac{(2x+4)^2}{4(x^2+4x+4+y^2)^{(3/2)}} - \frac{2}{\sqrt{x^2+4x+4+y^2}} - \frac{y^2}{(x^2-4x+4+y^2)^{(3/2)}} + \frac{y^2}{(x^2+4x+4+y^2)^{(3/2)}}$$

$$Lab := -\frac{(2a-4)^2}{4(a^2-4a+4+b^2)^{(3/2)}} + \frac{2}{\sqrt{a^2-4a+4+b^2}} + \frac{(2a+4)^2}{4(a^2+4a+4+b^2)^{(3/2)}} - \frac{2}{\sqrt{a^2+4a+4+b^2}} - \frac{b^2}{(a^2-4a+4+b^2)^{(3/2)}} + \frac{b^2}{(a^2+4a+4+b^2)^{(3/2)}}$$

NOTE: Notice, the definition and derivatives of *expressions* were easy, but the evaluations were tedious. The *arrow definition* was slightly more complicated than the expression definition, but the derivatives were no more complicated and the evaluations were much easier.

3.1.4 Gradient and Hessian

The vector of first partial derivatives of a function f is called the gradient of f. The matrix of second partial derivatives of f is called it's Hessian. They may be computed using the commands `Gradient` and `Hessian` from the `VecCalc` package or their aliases `Grad` and `Hess`. By default, these commands produce an angle bracket row `Vector` and a `Matrix`, respectively.

Example: 3.4. Enter the function $f(x, y, z) = x^3y^4z^5$ in arrow notation and compute the gradient and hessian and evaluate them at the point $P = (a, b, c)$.

SOLUTION: We enter the function and compute the gradient and hessian:

```
> f:=<x,y,z> &-> (x^3 * y^4 * z^5);
```

$$f := (x,\ y,\ z) \to x^3\, y^4\, z^5$$

```
> delf:=Grad(f); whattype(%);
```

$$delf := [(x,\ y,\ z) \to 3\,x^2\,y^4\,z^5,\ (x,\ y,\ z) \to 4\,x^3\,y^3\,z^5,\ (x,\ y,\ z) \to 5\,x^3\,y^4\,z^4]$$

$$Vector_{row}$$

```
> Hf:=Hess(f); whattype(%);
```

$$Hf := \begin{bmatrix} (x,\ y,\ z) \to 6\,x\,y^4\,z^5 & (x,\ y,\ z) \to 12\,x^2\,y^3\,z^5 & (x,\ y,\ z) \to 15\,x^2\,y^4\,z^4 \\ (x,\ y,\ z) \to 12\,x^2\,y^3\,z^5 & (x,\ y,\ z) \to 12\,x^3\,y^2\,z^5 & (x,\ y,\ z) \to 20\,x^3\,y^3\,z^4 \\ (x,\ y,\ z) \to 15\,x^2\,y^4\,z^4 & (x,\ y,\ z) \to 20\,x^3\,y^3\,z^4 & (x,\ y,\ z) \to 20\,x^3\,y^4\,z^3 \end{bmatrix}$$

$$Matrix$$

Since the results involve angle brackets, we evaluate them using `evalFunction` or `&@`:

```
> P:=<a|b|c>;
```

$$P := [a,\ b,\ c]$$

```
> delf&@P;
```

$$[3\,a^2\,b^4\,c^5,\ 4\,a^3\,b^3\,c^5,\ 5\,a^3\,b^4\,c^4]$$

```
> Hf&@P;
```

$$\begin{bmatrix} 6\,a\,b^4\,c^5 & 12\,a^2\,b^3\,c^5 & 15\,a^2\,b^4\,c^4 \\ 12\,a^2\,b^3\,c^5 & 12\,a^3\,b^2\,c^5 & 20\,a^3\,b^3\,c^4 \\ 15\,a^2\,b^4\,c^4 & 20\,a^3\,b^3\,c^4 & 20\,a^3\,b^4\,c^3 \end{bmatrix}$$

NOTE: A matrix can be entered using angle brackets. For example, the matrix $M = \begin{pmatrix} 2 & 4 & 6 \\ 1 & 3 & 5 \end{pmatrix}$ is entered as

```
> M:=<<2|4|6>,<1|3|5>>;
```

$$M := \begin{bmatrix} 2 & 4 & 6 \\ 1 & 3 & 5 \end{bmatrix}$$

Notice that each inner vector is a row of the matrix. By interchanging the commas and vertical bars, the inner vectors become the columns of the matrix.

```
> M:=<<2,1>|<4,3>|<6,5>>;
```

$$M := \begin{bmatrix} 2 & 4 & 6 \\ 1 & 3 & 5 \end{bmatrix}$$

If you prefer `Grad` and `Hess` to produce a `list` and a `list` of `list`'s, then first execute the commands:

```
> OutputVectorType:=list:  OutputMatrixType:=list:
```

Then we have

```
> delf:=Grad(f); whattype(%);
```

$$delf := [(x, y, z) \to 3\,x^2\,y^4\,z^5,\ (x, y, z) \to 4\,x^3\,y^3\,z^5,\ (x, y, z) \to 5\,x^3\,y^4\,z^4]$$

$$list$$

```
> Hf:=Hess(f); whattype(%);
```

$$Hf := [[(x, y, z) \to 6\,x\,y^4\,z^5,\ (x, y, z) \to 12\,x^2\,y^3\,z^5,\ (x, y, z) \to 15\,x^2\,y^4\,z^4],$$
$$[(x, y, z) \to 12\,x^2\,y^3\,z^5,\ (x, y, z) \to 12\,x^3\,y^2\,z^5,\ (x, y, z) \to 20\,x^3\,y^3\,z^4],$$
$$[(x, y, z) \to 15\,x^2\,y^4\,z^4,\ (x, y, z) \to 20\,x^3\,y^3\,z^4,\ (x, y, z) \to 20\,x^3\,y^4\,z^3]]$$

$$list$$

which can be evaluated more easily as

```
> delf(a,b,c);
```

$$[3\,a^2\,b^4\,c^5,\ 4\,a^3\,b^3\,c^5,\ 5\,a^3\,b^4\,c^4]$$

```
> Hf(a,b,c);
```

$$[[6\,a\,b^4\,c^5,\ 12\,a^2\,b^3\,c^5,\ 15\,a^2\,b^4\,c^4],\ [12\,a^2\,b^3\,c^5,\ 12\,a^3\,b^2\,c^5,\ 20\,a^3\,b^3\,c^4],$$
$$[15\,a^2\,b^4\,c^4,\ 20\,a^3\,b^3\,c^4,\ 20\,a^3\,b^4\,c^3]]$$

althought the `&@` syntax will still work. To see the Hessian as a matrix, you then need to use:

```
> Matrix(Hf(a,b,c));
```

$$\begin{bmatrix} 6\,a\,b^4\,c^5 & 12\,a^2\,b^3\,c^5 & 15\,a^2\,b^4\,c^4 \\ 12\,a^2\,b^3\,c^5 & 12\,a^3\,b^2\,c^5 & 20\,a^3\,b^3\,c^4 \\ 15\,a^2\,b^4\,c^4 & 20\,a^3\,b^3\,c^4 & 20\,a^3\,b^4\,c^3 \end{bmatrix}$$

3.2 Applications

3.2.1 Tangent Plane to a Graph

Given a function $f(x, y)$, the equation of the plane tangent to the graph $z = f(x, y)$ at the point where $(x, y) = (a, b)$ is:

$$z = f_{\tan}(x, y) = f(a, b) + f_x(a, b)(x - a) + f_y(a, b)(y - b).$$

Example: 3.5. Find the equation of the plane tangent to the ellipsoid $z = \sqrt{4 - \frac{x^2}{8} - \frac{y^2}{9}}$ at the point $(4,3)$. Then plot the upper half of the ellipsoid and the tangent plane.

SOLUTION: Enter the function and compute the partial derivatives:

```
> f:=<x,y> &-> sqrt(4-x^2/8-y^2/9);
```

$$f := (x, y) \rightarrow \frac{1}{12}\sqrt{576 - 18\,x^2 - 16\,y^2}$$

```
> fx:=D[1](f);
```

$$fx := (x, y) \rightarrow -\frac{3}{2}\frac{x}{\sqrt{576 - 18\,x^2 - 16\,y^2}}$$

```
> fy:=D[2](f);
```

$$fy := (x, y) \rightarrow -\frac{4}{3}\frac{y}{\sqrt{576 - 18\,x^2 - 16\,y^2}}$$

Define the function for the tangent plane:

```
> ftan:=<x,y> &-> simplify(f(4,3) + fx(4,3)*(x-4) + fy(4,3)*(y-3));
```

$$ftan := (x, y) \rightarrow 4 - \frac{1}{2}x - \frac{1}{3}y$$

So the tangent plane is:

```
> z = ftan(x,y);
```

$$z = 4 - \frac{x}{2} - \frac{y}{3}$$

Finally, plot the ellipsoid and the tangent plane:

```
> plot3d({f(x,y), ftan(x,y)}, x=-sqrt(32)..sqrt(32), y=-6..6,
> axes=normal, orientation=[-45,85]);
```

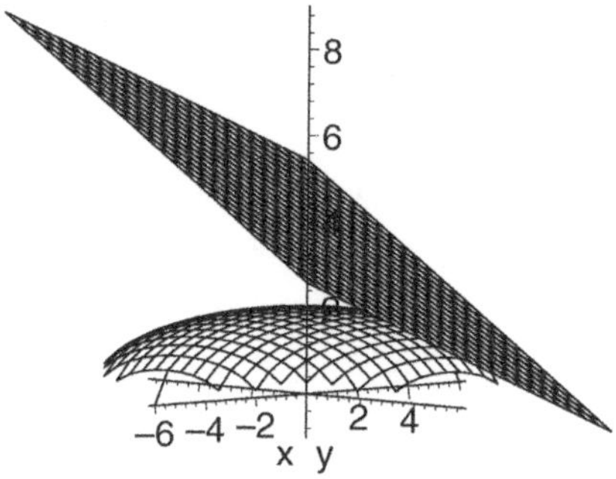

3.2.2 Differentials and the Linear Approximation

For points near the point of tangency, the tangent plane is close to the graph of a function. Hence, the function $f_{\tan}(x, y) = f(a, b) + f_x(a, b)(x - a) + f_y(a, b)(y - b)$ which defines the tangent plane as $z = f_{\tan}(x, y)$ is called the linear approximation to the function f at (a, b) and may be used to approximate the function $z = f(x, y)$ near the point of tangency (a, b).

If a point moves from (a, b) to $(x, y) = (a+\Delta x, b+\Delta y)$, then the differentials of the independent coordinates, x and y, are simply the changes in the coordinates:

$$dx = \Delta x = x - a \qquad \text{and} \qquad dy = \Delta y = y - b.$$

For the dependent coordinate, z, the change is:

$$\Delta z = f(x, y) - f(a, b) = f(a + \Delta x, b + \Delta y) - f(a, b),$$

while the differential is the change in the linear approximation:

$$\begin{aligned} dz &= f_{\tan}(x, y) - f_{\tan}(a, b) = f_{\tan}(a + \Delta x, b + \Delta y) - f(a, b) \\ &= f_x(a, b)\Delta x + f_y(a, b)\Delta y = f_x(a, b)\, dx + f_y(a, b)\, dy \end{aligned}$$

The linear approximation says that for points near (a, b), the tangent function $f_{\tan}(x, y)$ is close to the function $f(x, y)$, and therefore, that the differential dz is close to the change Δz.

Example: 3.6. A cylindrical can has radius $r = 4$ cm, height $h = 10$ cm and is made from aluminum which is .02 cm thick. Use differentials to estimate the volume of aluminum needed to make the can. Include the sides, top and bottom.

SOLUTION: The volume of a cylinder is

```
> V:=(r,h) -> Pi*r^2*h;
```

$$V := (r, h) \to \pi\, r^2\, h$$

To find the volume of the aluminum, we compare the volume inside the can to the volume including the metal. By the linear approximation, the change in the volume is approximately the differential of the volume. Thus,

$$\Delta V \approx dV = \frac{\partial V}{\partial r}(4, 10)\, dr + \frac{\partial V}{\partial h}(4, 10)\, dh$$

The partial derivatives are

```
> Vr:= D[1](V);
```

$$Vr := (r, h) \to 2\,\pi\, r\, h$$

```
> Vh:= D[2](V);
```

$$Vh := (r, h) \to \pi\, r^2$$

The change in the radius is the thickness of the aluminum:

```
> dr:= .02:
```

The change in the height is twice the thickness of the aluminum since there is a top and a bottom:

```
> dh:= .04:
```

Thus the volume of aluminum is approximately the differential of the volume:

```
> dV:=evalf(Vr(4,10)*dr + Vh(4,10)*dh);
```

$$dV := 7.037167545$$

Example: 3.7. Consider the surface in $\mathbb{R}^3$ given by the equation

$$F(x,y,z) = z^{10} + x^2y^2z^8 + x^4z^6 + y^4z^4 + x^2y^2z^2 + x^2 + 2y^2 = 8$$

(a) Verify that the point $(1,1,1)$ is on the surface.

Notice that this equation implicitly defines z as a function of x and y in the neighborhood of the point $(1,1,1)$. So we can write $z = f(x,y)$ where $f(1,1) = 1$.

(b) Use implicit differentiation to compute $\frac{\partial f}{\partial x}(1,1)$ and $\frac{\partial f}{\partial y}(1,1)$.

(c) Find the equation of the plane tangent the graph $z = f(x,y)$ at $(1,1)$.

(d) Use the linear approximation to $f(x,y)$ at $(x,y) = (1,1)$ to estimate $f(1.03, .98)$.

(e) Use `implicitplot3d` to plot the surface $F(x,y,z) = 8$. Use `plot3d` to plot the tangent plane at $(1,1,1)$. Then use `display` to put the two plots together.

SOLUTION: (a) We define the function F:

```
> F := MF([x,y,z], z^10 + x^2*y^2*z^8 + x^4*z^6 + y^4*z^4 + x^2*y^2*z^2
> + x^2 + 2*y^2);
```

$$F := (x,\, y,\, z) \to z^{10} + x^2\,y^2\,z^8 + x^4\,z^6 + y^4\,z^4 + x^2\,y^2\,z^2 + x^2 + 2\,y^2$$

and evaluate at $(1,1,1)$ to check that $F(1,1,1) = 8$:

```
> F(1,1,1);
```

$$8$$

(b) We substitute $z = f(x,y)$ into F to obtain the equation which implicitly defines f:

```
> eq:= F(x,y,f(x,y)) = 8;
```

$$eq := \mathrm{f}(x,\, y)^{10} + x^2\,y^2\,\mathrm{f}(x,\, y)^8 + x^4\,\mathrm{f}(x,\, y)^6 + y^4\,\mathrm{f}(x,\, y)^4 + x^2\,y^2\,\mathrm{f}(x,\, y)^2 + x^2 + 2\,y^2 = 8$$

Then we differentiate with respect to x, solve for $\frac{\partial f}{\partial x}$ and substitute $(x,y,z) = (1,1,1)$.

```
> diff(eq,x);
```

$$\begin{aligned}&10\,\mathrm{f}(x,\, y)^9\,(\tfrac{\partial}{\partial x}\,\mathrm{f}(x,\, y)) + 2\,x\,y^2\,\mathrm{f}(x,\, y)^8 + 8\,x^2\,y^2\,\mathrm{f}(x,\, y)^7\,(\tfrac{\partial}{\partial x}\,\mathrm{f}(x,\, y)) + 4\,x^3\,\mathrm{f}(x,\, y)^6\\&+ 6\,x^4\,\mathrm{f}(x,\, y)^5\,(\tfrac{\partial}{\partial x}\,\mathrm{f}(x,\, y)) + 4\,y^4\,\mathrm{f}(x,\, y)^3\,(\tfrac{\partial}{\partial x}\,\mathrm{f}(x,\, y)) + 2\,x\,y^2\,\mathrm{f}(x,\, y)^2\\&+ 2\,x^2\,y^2\,\mathrm{f}(x,\, y)\,(\tfrac{\partial}{\partial x}\,\mathrm{f}(x,\, y)) + 2\,x = 0\end{aligned}$$

```
> fxsol:= solve(%, diff(f(x,y),x));
```

$$fxsol := -\frac{x\,(y^2\,\mathrm{f}(x,\, y)^8 + 2\,x^2\,\mathrm{f}(x,\, y)^6 + y^2\,\mathrm{f}(x,\, y)^2 + 1)}{\mathrm{f}(x,\, y)\,(5\,\mathrm{f}(x,\, y)^8 + 4\,\mathrm{f}(x,\, y)^6\,x^2\,y^2 + 3\,\mathrm{f}(x,\, y)^4\,x^4 + 2\,\mathrm{f}(x,\, y)^2\,y^4 + x^2\,y^2)}$$

```
> fx:= MF([x,y,z], subs(f(x,y)=z,fxsol));
```

$$fx := (x,\, y,\, z) \to -\frac{x\,(y^2\,z^8 + 2\,x^2\,z^6 + y^2\,z^2 + 1)}{z\,(5\,z^8 + 4\,z^6\,x^2\,y^2 + 3\,z^4\,x^4 + 2\,z^2\,y^4 + x^2\,y^2)}$$

```
> fx0:= fx(1,1,1);
```

$$fx0 := \frac{-1}{3}$$

Likewise for y:

```
> diff(eq,y);
```

$$10\,f(x,y)^9\left(\frac{\partial}{\partial y}f(x,y)\right)+2x^2y\,f(x,y)^8+8x^2y^2f(x,y)^7\left(\frac{\partial}{\partial y}f(x,y)\right)+6x^4f(x,y)^5\left(\frac{\partial}{\partial y}f(x,y)\right)$$
$$+4y^3f(x,y)^4+4y^4f(x,y)^3\left(\frac{\partial}{\partial y}f(x,y)\right)+2x^2y\,f(x,y)^2+2x^2y^2f(x,y)\left(\frac{\partial}{\partial y}f(x,y)\right)$$
$$+4y=0$$

```
> fysol:= solve(%, diff(f(x,y),y));
```

$$fysol := -\frac{y\,(x^2f(x,y)^8+2y^2f(x,y)^4+x^2f(x,y)^2+2)}{f(x,y)\,(5f(x,y)^8+4f(x,y)^6x^2y^2+3f(x,y)^4x^4+2f(x,y)^2y^4+x^2y^2)}$$

```
> fy:= MF([x,y,z], subs(f(x,y)=z,fysol));
```

$$fy := (x,y,z)\to-\frac{y\,(x^2z^8+2y^2z^4+x^2z^2+2)}{z\,(5z^8+4z^6x^2y^2+3z^4x^4+2z^2y^4+x^2y^2)}$$

```
> fy0:= fy(1,1,1);
```

$$fy0 := \frac{-2}{5}$$

(c) We define the tangent function using the fact that $f(1,1)=1$:

```
> ftan:= MF([x,y], 1 + fx0*(x-1) + fy0*(y-1));
```

$$ftan := (x,y)\to\frac{26}{15}-\frac{1}{3}x-\frac{2}{5}y$$

Then the tangent plane is:

```
> z = ftan(x,y);
```

$$z=\frac{26}{15}-\frac{x}{3}-\frac{2y}{5}$$

(d) The linear approximation to f is just the tangent function. So we evaluate it at $(1.03, .98)$:

```
> ftan(1.03,.98);
```

$$0.9980000000$$

If you are curious, you can compare this result from the linear approximation with the result from the `fsolve` command:

```
> fsolve(F(1.03,.98,z)=8,z);
```

$$-0.9977857069,\ 0.9977857069$$

Pretty close.

(e) We plot of the equation $F(x,y,z)=8$ and save it as plotF:

```
> plotF:= implicitplot3d( F(x,y,z)=8, x=0..2, y=0..2, z=0..2):
```

Then we plot the tangent plane $z=f_{\tan}(x,y)$ and save it as plotFtan:

```
> plotFtan:= plot3d( ftan(x,y), x=0..2, y=0..2, color=gray):
```

Finally, we display the two plots together:

```
> display({plotF, plotFtan}, orientation=[30,105],
> scaling=constrained, axes=normal);
```

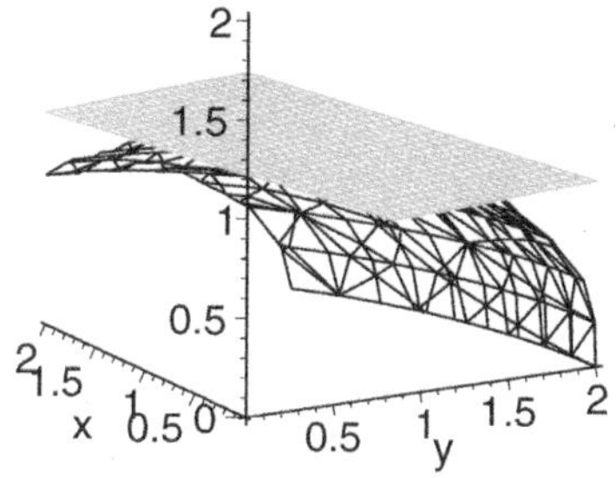

3.2.3 Taylor Polynomial Approximations

For a function of one variable, $f(x)$, the linear approximation may be improved by using the quadratic approximation or a higher order Taylor polynomial approximation. In general, the nth order Taylor approximation to $f(x)$ at $x = a$ is

$$f_n(x) = f(a) + f'(a)(x-a) + \frac{f''(a)}{2}(x-a)^2 + \cdots + \frac{f^{(n)}(a)}{n!}(x-a)^n$$

Using *Maple*, a Taylor polynomial may be computed using the `taylor` and `convert(..., polynom)` commands. For example, the 5th order Taylor polynomial for $\ln(x)$ at $x = 2$ is

```
> t5:=taylor(ln(x),x=2,6);
```

$$t5 := \ln(2) + \frac{1}{2}(x-2) - \frac{1}{8}(x-2)^2 + \frac{1}{24}(x-2)^3 - \frac{1}{64}(x-2)^4 + \frac{1}{160}(x-2)^5 + \mathrm{O}((x-2)^6)$$

```
> convert(t5,polynom);
```

$$\ln(2) + \frac{x}{2} - 1 - \frac{(x-2)^2}{8} + \frac{(x-2)^3}{24} - \frac{(x-2)^4}{64} + \frac{(x-2)^5}{160}$$

NOTE: The last parameter to `taylor` is the integer one greater than the order of the Taylor polynomial. In fact, this parameter is the order of the error term, shown as $O(x^6)$. The `convert` command strips off the order term.

Similarly, for a function of several variables, $f(x_1, x_2, \ldots, x_k)$, the linear approximation may be improved by using the quadratic approximation or a higher order Taylor polynomial approximation. In general, the nth order Taylor approximation to $f(x_1, x_2, \ldots, x_k)$ at $(x_1, x_2, \ldots, x_k) = (a_1, a_2, \ldots, a_k)$ is

$$f_n(\vec{x}) = f(\vec{a}) + \sum_{i=1}^{k} \frac{\partial f}{\partial x_i}(\vec{a})(x_i - a_i) + \frac{1}{2}\sum_{i=1}^{k}\sum_{j=1}^{k} \frac{\partial^2 f}{\partial x_i \partial x_j}(\vec{a})(x_i - a_i)(x_j - a_j)$$

$$+ \cdots + \frac{1}{n!}\sum_{i_1=1}^{k} \cdots \sum_{i_n=1}^{k} \frac{\partial^n f}{\partial x_{i_1} \cdots \partial x_{i_n}}(\vec{a})(x_{i_1} - a_{i_1}) \cdots (x_{i_n} - a_{i_n})$$

Using *Maple*, a multi-variable Taylor polynomial may be computed using the `mtaylor` command.

For example, the 3rd order Taylor polynomial for $\sqrt{x^2+y^2}$ at $(x,y)=(4,3)$ is

```
> t3:=mtaylor(sqrt(x^2 +y^2), [x=4,y=3], 4);
```

$$t3 := \frac{4\,x}{5}+\frac{3\,y}{5}+\frac{9\,(x-4)^2}{250}-\frac{12\,(x-4)\,(y-3)}{125}+\frac{8\,(y-3)^2}{125}-\frac{18\,(x-4)^3}{3125}$$
$$+\frac{69\,(y-3)\,(x-4)^2}{6250}+\frac{4\,(y-3)^2\,(x-4)}{3125}-\frac{24\,(y-3)^3}{3125}$$

NOTE: The last parameter to `mtaylor` is the integer one greater than the order of the Taylor polynomial.The `mtaylor` command does not produce an order term. So you do not need the `convert` command.

Example: 3.8. Find the Taylor polynomials for $f(x,y)=\sin(x)\cos(y)$ about $(x,y)=(0,0)$ of orders 3, 11 and 19. Then display the ordinary plots and contour plots for the original function and each of the Taylor polynomials.

SOLUTION: Enter the function:

```
> f:=sin(x)*cos(y);
```

$$f := \sin(x)\cos(y)$$

Then compute the Taylor polynomials: (The output is so long that we will only display the first polynomial.)

```
> f3:=mtaylor(f,[x=0,y=0],4);
```

$$f3 := x-\frac{1}{2}\,y^2\,x-\frac{1}{6}\,x^3$$

```
> f11:=mtaylor(f,[x=0,y=0],12):
> f19:=mtaylor(f,[x=0,y=0],20):
```

The command `nops` will count the number of terms in a sum. In particular,

```
> nops(f3), nops(f11), nops(f19);
```

$$3,\ 21,\ 55$$

So f_3 has 3 terms, f_{11} has 21 terms and f_{19} has 55 terms.

Next, compute the ordinary plots of the function and of the Taylor polynomials:

```
> plot3d(f, x=-2*Pi..2*Pi, y=-2*Pi..2*Pi, orientation=[15,30],
> view=-2..2);
```

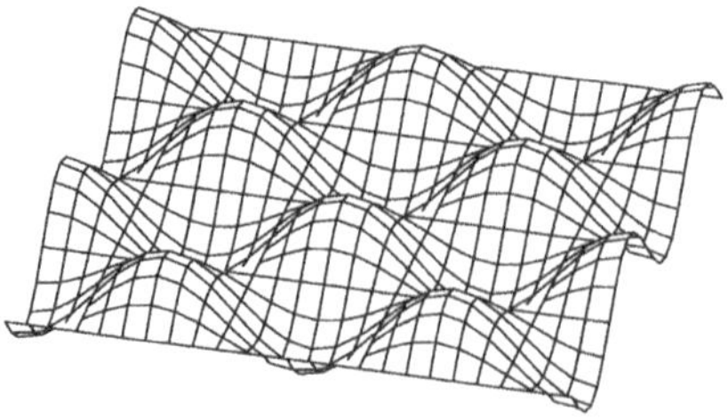

```
> plot3d(f3,  x=-2*Pi..2*Pi, y=-2*Pi..2*Pi, orientation=[15,30],
> view=-2..2);
```

```
> plot3d(f11, x=-2*Pi..2*Pi, y=-2*Pi..2*Pi, orientation=[15,30],
> view=-2..2);
```

```
> plot3d(f19, x=-2*Pi..2*Pi, y=-2*Pi..2*Pi, orientation=[15,30],
> view=-2..2);
```

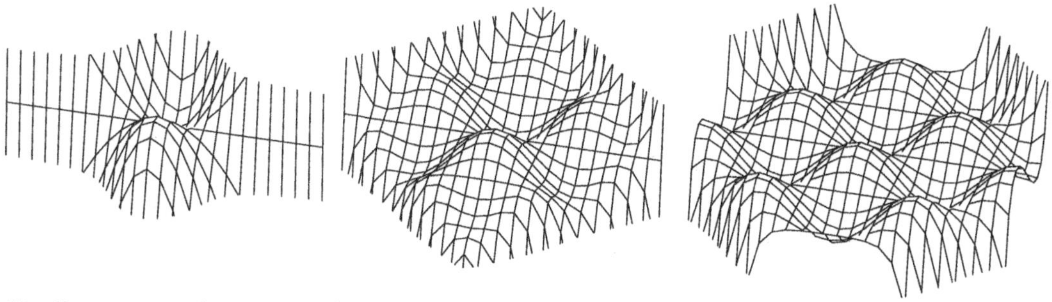

Finally, compute the contour plots of the function and of the Taylor polynomials:

```
> contourplot3d(f,  x=-2*Pi..2*Pi, y=-2*Pi..2*Pi, orientation=[-90,0],
> view=-2..2);
```

```
> contourplot3d(f3,  x=-2*Pi..2*Pi, y=-2*Pi..2*Pi, orientation=[-90,0],
> view=-2..2);
```

```
> contourplot3d(f11, x=-2*Pi..2*Pi, y=-2*Pi..2*Pi, orientation=[-90,0],
> view=-2..2);
```

```
> contourplot3d(f19, x=-2*Pi..2*Pi, y=-2*Pi..2*Pi, orientation=[-90,0],
> view=-2..2);
```

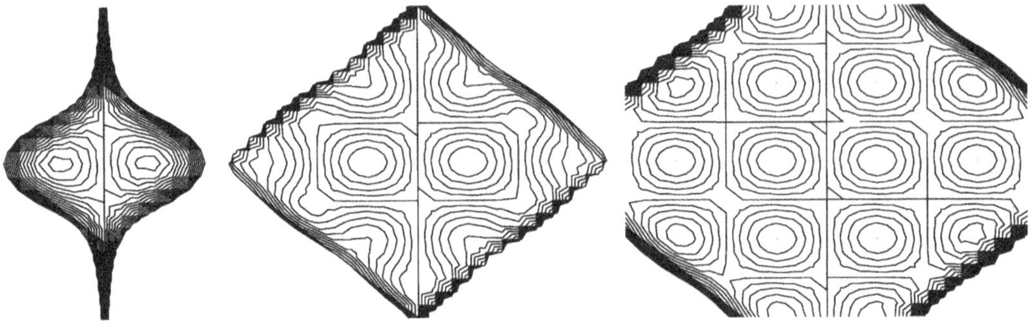

Notice that the Taylor polynomials become better approximations to the function as the number of terms increases.

3.2.4 Chain rule

Suppose z is a function of x and y, i.e. $z = z(x, y)$, while x and y are functions of t, i.e. $x = x(t)$ and $y = y(t)$. Then z may also be regarded as a function of t through the composition $z = z(x(t), y(t))$ and its derivative may be compute by using the chain rule:

$$\frac{dz}{dt} = \frac{\partial z}{\partial x}(x(t), y(t))\frac{dx}{dt} + \frac{\partial z}{\partial y}(x(t), y(t))\frac{dy}{dt}$$

Similarly, if z is a function of x and y, while x and y are functions of s and t then the chain rule formulas are:

$$\frac{\partial z}{\partial s} = \frac{\partial z}{\partial x}(x(s,t), y(s,t))\frac{\partial x}{\partial s} + \frac{\partial z}{\partial y}(x(s,t), y(s,t))\frac{\partial y}{\partial s}$$

and

$$\frac{\partial z}{\partial t} = \frac{\partial z}{\partial x}(x(s,t), y(s,t))\frac{\partial x}{\partial t} + \frac{\partial z}{\partial y}(x(s,t), y(s,t))\frac{\partial y}{\partial t}$$

These formulas may be generalized to larger numbers of variables.

Example: 3.9. A starship is travelling through a high temperature plasma field. Its shields are capable of withstanding very high temperatures but can only adjust to these temperatures at a rate no greater than 25°C per second. Assume that the temperature distribution in the plasma is the Gaussian distribution

$$T = 12500°\text{C}\ \mathrm{e}^{-(x^2+y^2+z^2)/10000}$$

and the starship is travelling along the parabolic curve $y = x^2 - 100$, $z = 0$ as a function of time according to

$$(x, y, z) = \vec{r}(t) = \left(\frac{\operatorname{arcsinh}(t)}{2}, \frac{\operatorname{arcsinh}(t)^2}{4} - 100, 0\right)$$

where all distances are given in light-seconds and time is given in seconds. Plot the absolute value of the expected rate of change of temperature to ensure that it is never greater than 25°C per second.

SOLUTION: We enter the temperature function and the parametrized curve:

```
> T:=MF(<x,y,z>, 12500*exp(-(x^2+y^2+z^2)/10000) );
```

$$T := (x,\, y,\, z) \to 12500\, e^{(-1/10000\, x^2 - 1/10000\, y^2 - 1/10000\, z^2)}$$

```
> r:=MF(t,<arcsinh(t)/2| arcsinh(t)^2/4-100| 0>);
```

$$r := \left[t \to \frac{1}{2}\operatorname{arcsinh}(t),\, t \to \frac{1}{4}\operatorname{arcsinh}(t)^2 - 100,\, t \to 0\right]$$

We will find the derivative in three ways.

Method 1 We form the composition $T(\vec{r}(t))$:

```
> Tr:=T &@ (r &@ t);
```

$$Tr := 12500\, e^{(-1/40000 \operatorname{arcsinh}(t)^2 - 1/10000\,(1/4 \operatorname{arcsinh}(t)^2 - 100)^2)}$$

NOTE: The parentheses guarantee that `r` is first evaluated at `t`, then `T` is evaluated at `r(t)`.

Then we take the derivative:

```
> DTr1:=diff(Tr,t);
```

$$DTr1 := 12500 \left(-\frac{1}{20000}\frac{\operatorname{arcsinh}(t)}{\sqrt{1+t^2}} - \frac{1}{10000}\frac{(\frac{1}{4}\operatorname{arcsinh}(t)^2 - 100)\operatorname{arcsinh}(t)}{\sqrt{1+t^2}} \right)$$
$$e^{(-1/40000 \operatorname{arcsinh}(t)^2 - 1/10000\,(1/4 \operatorname{arcsinh}(t)^2 - 100)^2)}$$

Method 2 The chain rule says

$$\frac{dT}{dt} = \frac{\partial T}{\partial x}(\vec{r}(t))\frac{dx}{dt} + \frac{\partial T}{\partial y}(\vec{r}(t))\frac{dy}{dt} + \frac{\partial T}{\partial z}(\vec{r}(t))\frac{dz}{dt}$$

So we compute the partial derivatives of T and the t derivatives of x, y and z and plug into the chain rule:

```
> Tx:=D[1](T); Ty:=D[2](T); Tz:=D[3](T);
```

$$Tx := (x,\, y,\, z) \to -\frac{5}{2}\, x\, e^{(-1/10000\, x^2 - 1/10000\, y^2 - 1/10000\, z^2)}$$

$$Ty := (x,\, y,\, z) \to -\frac{5}{2}\, y\, e^{(-1/10000\, x^2 - 1/10000\, y^2 - 1/10000\, z^2)}$$

$$Tz := (x,\, y,\, z) \to -\frac{5}{2}\, z\, e^{(-1/10000\, x^2 - 1/10000\, y^2 - 1/10000\, z^2)}$$

```
> Dx:=D(r[1]); Dy:=D(r[2]); Dz:=D(r[3]);
```

$$Dx := t \to \frac{1}{2}\frac{1}{\sqrt{1+t^2}}$$

$$Dy := t \to \frac{1}{2}\frac{\operatorname{arcsinh}(t)}{\sqrt{1+t^2}}$$

$$Dz := 0$$

```
> DTr2:=Tx&@(r&@t)*Dx(t) + Ty&@(r&@t)*Dy(t) + Tz&@(r&@t)*Dz(t);
```

$$DTr2 := -\frac{5}{8}\,\frac{\operatorname{arcsinh}(t)\, e^{(-1/40000 \operatorname{arcsinh}(t)^2 - 1/10000\,(1/4 \operatorname{arcsinh}(t)^2 - 100)^2)}}{\sqrt{1+t^2}}$$
$$-\frac{5}{4}\,\frac{(\frac{1}{4}\operatorname{arcsinh}(t)^2 - 100)\, e^{(-1/40000 \operatorname{arcsinh}(t)^2 - 1/10000\,(1/4 \operatorname{arcsinh}(t)^2 - 100)^2)} \operatorname{arcsinh}(t)}{\sqrt{1+t^2}}$$

Although the answers look different, we can check they are the same:

```
> simplify(DTr1-DTr2);
```

$$0$$

Method 3 Notice that the chain rule formula may be written as the dot product

$$\frac{dT}{dt} = \vec{\nabla}T(\vec{r}(t)) \cdot \vec{v}(t)$$

of the gradient $\vec{\nabla}T$ of the temperature T evaluated on the curve $\vec{r}(t)$ and the velocity $\vec{v}(t)$ of the curve $\vec{r}(t)$. So we compute the gradient of T and the velocity of $\vec{r}$ and take the dot product:

```
> delT:=Grad(T)&@(r&@t);
```

$$delT := \Big[-\frac{5}{4}\operatorname{arcsinh}(t)\, e^{(-1/40000 \operatorname{arcsinh}(t)^2 - 1/10000\,(1/4 \operatorname{arcsinh}(t)^2 - 100)^2)},$$
$$-\frac{5}{2}\,(\frac{1}{4}\operatorname{arcsinh}(t)^2 - 100)\, e^{(-1/40000 \operatorname{arcsinh}(t)^2 - 1/10000\,(1/4 \operatorname{arcsinh}(t)^2 - 100)^2)},\ 0\Big]$$

```
> v:=D(r)&@t;
```

$$v := \left[\frac{1}{2\sqrt{1+t^2}},\ \frac{1}{2}\,\frac{\operatorname{arcsinh}(t)}{\sqrt{1+t^2}},\ 0\right]$$

```
> DTr3:=delT &.  v;
```

$$DTr3 := -\frac{5}{16}\,\frac{\operatorname{arcsinh}(t)\, e^{(\frac{199}{40000}\operatorname{arcsinh}(t)^2 - 1/160000 \operatorname{arcsinh}(t)^4 - 1)}\,(-398 + \operatorname{arcsinh}(t)^2)}{\sqrt{1+t^2}}$$

Again, we can check the answers are the same:

```
> simplify(DTr1-DTr3);
```

$$0$$

Now that we have computed the rate of change of the temperature, we can plot the absolute value of this rate and the horizontal line at 25, to be sure that the temperature is never changing faster than 25°C per second.

```
> plot ({abs(DTr1),25}, t=-10..10);
```

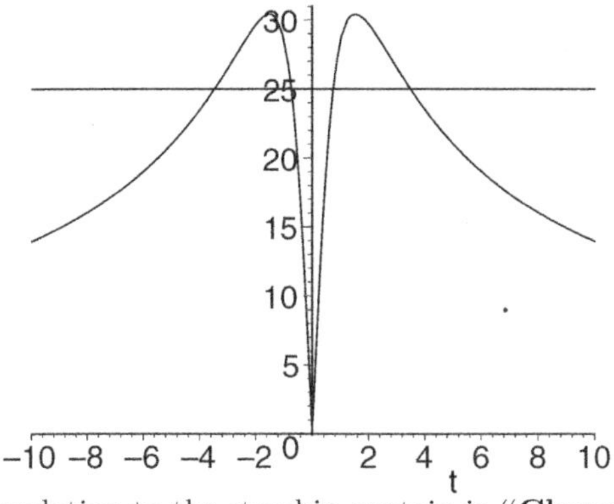

So our recommendation to the starship captain is **"Change course or put on the breaks!"**

The first method was the fastest, but it can only be used if you have explicit formulas for the outer function, here $T(x, y, z)$, and the inner functions, here $\vec{r}(t) = (x(t), y(t), z(t))$. This will not be the case in the next example. The second and third methods are essentially the same computation but the third method is independent of the dimension of space and so is faster when there are more intermediate variables. The third method will be generalized to the derivative along a curve and the directional derivative in the next subsection.

Example: 3.10. In example 3.7, we saw that the surface given by the equation

$$F(x, y, z) = z^{10} + x^2y^2z^8 + x^4z^6 + y^4z^4 + x^2y^2z^2 + x^2 + 2y^2 = 8$$

defines z as a function of x and y in the neighborhood of the point $(1, 1, 1)$. So we were able to write $z = f(x, y)$ where $f(1, 1) = 1$. Then we computed $\frac{\partial f}{\partial x}(1, 1)$ and $\frac{\partial f}{\partial y}(1, 1)$. We now transform to polar (or cylindrical) coordinates using the equations

$$x = r\cos(\theta) \qquad y = r\sin(\theta).$$

Then the point $(x, y) = (1, 1)$ has polar coordinates $(r, \theta) = (\sqrt{2}, \frac{\pi}{4})$ and the same surface $F(x, y, z) = 8$ defines z as a function of r and θ in the neighborhood of the point $(r, \theta, z) = (\sqrt{2}, \frac{\pi}{4}, 1)$. So we can also write $z = g(r, \theta)$ where $g(\sqrt{2}, \frac{\pi}{4}) = 1$. Use the chain rule to compute $\frac{\partial g}{\partial r}(\sqrt{2}, \frac{\pi}{4})$ and $\frac{\partial g}{\partial \theta}(\sqrt{2}, \frac{\pi}{4})$.

SOLUTION: In example 3.7, we entered the function F, used implicit differentiation to compute $\frac{\partial f}{\partial x}$ and $\frac{\partial f}{\partial y}$ and evaluated them at $(x, y) = (1, 1)$. The results were

```
> fx0, fy0;
```

$$\frac{-1}{3}, \frac{-2}{5}$$

We now enter the polar coordinate formulas,

```
> x0:=MF([r,theta], r*cos(theta)); y0:=MF([r,theta], r*sin(theta));
```

$$x0 := (r, \theta) \to r\cos(\theta)$$

$$y0 := (r, \theta) \to r\sin(\theta)$$

compute their partial derivatives and evaluate them at $(x, y) = (1, 1)$ which corresponds to $(r, \theta) = (\sqrt{2}, \frac{\pi}{4})$:

```
> xr:=D[1](x0); xr0:=xr(sqrt(2),Pi/4);
```

$$xr := (r, \theta) \to \cos(\theta)$$

$$xr0 := \frac{\sqrt{2}}{2}$$

```
> xtheta:=D[2](x0); xtheta0:=xtheta(sqrt(2),Pi/4);
```

$$xtheta := (r, \theta) \to -r\sin(\theta)$$

$$xtheta0 := -1$$

```
> yr:=D[1](y0); yr0:=yr(sqrt(2),Pi/4);
```

$$yr := (r, \theta) \to \sin(\theta)$$

$$yr0 := \frac{\sqrt{2}}{2}$$

```
> ytheta:=D[2](y0); ytheta0:=ytheta(sqrt(2),Pi/4);
```

$$ytheta := (r, \theta) \to r\cos(\theta)$$

$$ytheta0 := 1$$

Then the chain rule says

$$\frac{\partial g}{\partial r}(\sqrt{2}, \frac{\pi}{4}) = \frac{\partial f}{\partial x}(1,1)\frac{\partial x}{\partial r}(\sqrt{2}, \frac{\pi}{4}) + \frac{\partial f}{\partial y}(1,1)\frac{\partial y}{\partial r}(\sqrt{2}, \frac{\pi}{4})$$

and

$$\frac{\partial g}{\partial \theta}(\sqrt{2}, \frac{\pi}{4}) = \frac{\partial f}{\partial x}(1,1)\frac{\partial x}{\partial \theta}(\sqrt{2}, \frac{\pi}{4}) + \frac{\partial f}{\partial y}(1,1)\frac{\partial y}{\partial \theta}(\sqrt{2}, \frac{\pi}{4})$$

NOTE: The partial derivatives of g, x and y are evaluated at $(\sqrt{2}, \frac{\pi}{4})$ since g, x and y are functions of r and θ while the partial derivatives of f are evaluated at $(1, 1)$ since f is a function of x and y.

So we compute: The partial derivative $\frac{\partial g}{\partial r}(\sqrt{2}, \frac{\pi}{4})$ is:

```
> gr0:=fx0*xr0 + fy0*yr0;
```

$$gr0 := -\frac{11\sqrt{2}}{30}$$

and the partial derivative $\frac{\partial g}{\partial \theta}(\sqrt{2}, \frac{\pi}{4})$ is:

```
> gtheta0:=fx0*xtheta0 + fy0*ytheta0;
```

$$gtheta0 := \frac{-1}{15}$$

3.2.5 Derivatives along a Curve and Directional Derivatives

In example 3.9, we computed the time derivative of the temperature as felt by a starship as it moved through a plasma field along a specified curve.

In general, if a point moves through space along a specified curve $\vec{r}(t) = \big(x(t), y(t), z(t)\big)$ and a function $f(x, y, z)$ is defined throughout space, then the composition $f\big(\vec{r}(t)\big) = f\big(x(t), y(t), z(t)\big)$ is called the restriction of the function f to the curve $\vec{r}(t)$ or the value of the function f along the curve $\vec{r}(t)$. Then the derivative of the composition

$$\frac{df}{dt} = \frac{df\big(\vec{r}(t)\big)}{dt} = \frac{df\big(x(t), y(t), z(t)\big)}{dt}$$

is called the derivative of f along the curve $\vec{r}(t)$. By the chain rule, this may be written as

$$\frac{df}{dt} = \frac{\partial f}{\partial x}(\vec{r}(t))\frac{dx}{dt} + \frac{\partial f}{\partial y}(\vec{r}(t))\frac{dy}{dt} + \frac{\partial f}{\partial z}(\vec{r}(t))\frac{dz}{dt} = \vec{\nabla} f(\vec{r}(t)) \cdot \vec{v}(t)$$

In the last step, the derivative of f along the curve $\vec{r}(t)$ has been written as the dot product of the gradient $\vec{\nabla} f$ evaluated on the curve $\vec{r}(t)$ and the velocity $\vec{v}(t)$ of the curve. It is important to remember that the derivative along a curve can be computed either by using the chain rule or by using its definition as the derivative of the composition.

More generally, if f is a function defined throughout space and $\vec{v}$ is a vector located at a point $\vec{x}$, then the derivative of f along the vector $\vec{v}$ at the point $\vec{x}$ is defined to be

$$\vec{\nabla}_{\vec{v}} f = \vec{v} \cdot \vec{\nabla} f(\vec{x}).$$

So, the derivative of f along a curve $\vec{r}(t)$ is the same as the derivative of f along its velocity vector $\vec{v}(t)$:

$$\frac{df}{dt} = \vec{v}(t) \cdot \vec{\nabla} f(\vec{r}(t)) = \vec{\nabla}_{\vec{v}} f$$

And conversely, the derivative of f along the vector $\vec{v}$ at the point $\vec{x}$ is the same as the derivative of f along any curve $\vec{r}(t)$ which passes through $\vec{x}$ and has velocity $\vec{v}$ there.

Further, as a special case, if $\hat{u}$ is a unit vector, then the derivative of f along the vector $\hat{u}$, namely $\vec{\nabla}_{\hat{u}} f$, is also called the directional derivative of f in the direction $\hat{u}$. In particular, if $\vec{v}$ is any vector, then the directional derivative of f in *the direction of* the vector $\vec{v}$ is $\vec{\nabla}_{\hat{v}} f$ where $\hat{v} = \dfrac{\vec{v}}{|\vec{v}|}$ is the unit vector in the direction of $\vec{v}$.

Example: 3.11. Consider the function $f = x^2 y^3 z^4$. Compute each of the following:

(a) The gradient of f at a general point and at the point $\vec{P} = (2, 4, 8)$.

(b) The derivative of f along the vector $\vec{v} = (1, 4, 12)$ at the point $\vec{P} = (2, 4, 8)$. (Notice that you do not need to know anything about a curve to compute this derivative.)

(c) The directional derivative of f in the direction of the vector $\vec{v} = (1, 4, 12)$ at the point $\vec{P} = (2, 4, 8)$.

(d) The derivative of f along the curve $\vec{r} = (t, t^2, t^3)$ at time $t = 2$. Compute this in two ways.

(e) The derivative of f along the curve $\vec{R} = (2T, 4T^2, 8T^3)$ at time $T = 1$. Comparing the curves $\vec{r}$ and $\vec{R}$, what can you say about their paths, their speeds and their velocities and the derivatives of f along the two curves.

(f) The derivative of f along the line $\vec{X} = (2 + u, 4 + 4u, 8 + 12u)$ at time $u = 0$. Comparing the curves $\vec{r}$ and $\vec{X}$, what can you say about their paths, their speeds and their velocities at the point $\vec{P} = (2, 4, 8)$ and the derivatives of f along the two curves at the point $\vec{P} = (2, 4, 8)$.

SOLUTION: We first enter the function:

```
> f:=[x,y,z] &-> (x^2*y^3*z^4);
```

$$f := (x, y, z) \to x^2 y^3 z^4$$

(a) We compute the gradient and evaluate it at $\vec{P} = (2, 4, 8)$:

```
> delf:=Grad(f);
```

$$delf := [(x, y, z) \to 2\,x\,y^3\,z^4,\ (x, y, z) \to 3\,x^2\,y^2\,z^4,\ (x, y, z) \to 4\,x^2\,y^3\,z^3]$$

```
> delfP:=delf&@ <2,4,8>;
```

$$delfP := [1048576,\ 786432,\ 524288]$$

(b) The derivative of f along the vector $\vec{v} = (1, 4, 12)$ at $\vec{P} = (2, 4, 8)$ is $\vec{\nabla}_{\vec{v}} f = \vec{v} \cdot \vec{\nabla} f(\vec{P})$:

```
> v:=<1|4|12>:  vdelfP:=v &.  delfP;
```

$$vdelfP := 10485760$$

(c) The unit vector in the direction of $\vec{v}$ is

```
> vhat:=v/Length(v);
```

$$vhat := \left[\frac{\sqrt{161}}{161},\ \frac{4\sqrt{161}}{161},\ \frac{12\sqrt{161}}{161}\right]$$

So the directional derivative of f in the direction of the vector $\vec{v}$ at the point $\vec{P} = (2, 4, 8)$ is:

```
> fvhat:= vhat &.  delfP;
```

$$fvhat := \frac{10485760\sqrt{161}}{161}$$

(d) We enter the curve $\vec{r}(t)$, differentiate the composition $f(\vec{r}(t))$ and evaluate at $t = 2$:

```
> r:= MF(t, <t| t^2| t^3>);
```

$$r := [t \to t,\ t \to t^2,\ t \to t^3]$$

```
> Dfr:= diff(f&@r&@t, t);
```

$$Dfr := 20\,t^{19}$$

```
> subs(t=2, Dfr);
```

$$10485760$$

Alternatively, we compute the velocity $\vec{v}(t)$ and the dot product $\vec{v}(t) \cdot \vec{\nabla} f(\vec{r}(t))$ and evaluate at $t = 2$:

```
> v:=D(r);
```

$$v := [1,\ t \to 2\,t,\ t \to 3\,t^2]$$

```
> v&@t;
```

$$[1,\ 2\,t,\ 3\,t^2]$$

```
> vdelfr:=v&@t &.  (delf&@(r&@t));
```

$$vdelfr := 20\,t^{19}$$

```
> subs(t=2, vdelfr);
```

$$10485760$$

NOTE: This is the same answer as in (b) since $\vec{v}(2) = (1, 4, 12)$ and $\vec{r}(2) = (2, 4, 8)$.

(e) We enter the curve $\vec{R}(T)$ and compute the velocity:

```
> R:= MF(T, <2*T| 4*T^2| 8*T^3>);
```

$$R := [T \to 2\,T,\ T \to 4\,T^2,\ T \to 8\,T^3]$$

```
> V:=D(R);
```

$$V := [2,\ T \to 8\,T,\ T \to 24\,T^2]$$

Then we compute the dot product $\vec{V}(T) \cdot \vec{\nabla} f(\vec{R}(T))$ and evaluate at $T = 1$:

```
> VdelfR:=V&@T &.  (delf&@(R&@T));
```

$$VdelfR := 20971520\,T^{19}$$

```
> subs(T=1, VdelfR);
```

$$20971520$$

The curves $\vec{r}(t)$ and $\vec{R}(T)$ follow the same path but they are parametrized differently. The parameters are related by $t = 2T$ since $\vec{R}(T) = \vec{r}(2T)$. Hence the velocities are related by

$$\vec{V}(T) = \frac{d\vec{R}(T)}{dT} = \frac{d\vec{r}(2T)}{dT} = 2\frac{d\vec{r}}{dt} = 2\vec{v}(2T).$$

Hence the speed for $\vec{R}(T)$ is twice the speed for $\vec{r}(t)$ and the derivative of f along $\vec{R}(T)$ is twice the derivative along $\vec{r}(t)$.

(f) This time we enter the curve $\vec{X}(u)$ and compute the position and velocity at time $u = 0$:

```
> X:= MF(u, <2+u| 4+4*u| 8+12*u>);
```

$$X := [u \to 2 + u,\ u \to 4 + 4\,u,\ u \to 8 + 12\,u]$$

```
> X&@0;
```

$$[2,\ 4,\ 8]$$

```
> v:=D(X);
```

$$v := [1,\ 4,\ 12]$$

```
> v&@0;
```

$$[1,\ 4,\ 12]$$

So the derivative along the curve is:

```
> vdelfX:= v&@0 &.  (delf&@(X&@0));
```

$$vdelfX := 10485760$$

The two curves $\vec{r}(t)$ and $\vec{X}(u)$ are not the same but they both pass through the point $(2, 4, 8)$ and have velocity $(1, 4, 12)$ there. In other words, the two curves are tangent at this point and have the same speed there. Hence, the derivative along the curve must be the same.

3.2.6 Interpretation of the Gradient

The gradient of a function f satisfies 4 properties:

1. $\vec{\nabla} f$ points in the direction of maximum increase of the function f.

2. $\left|\vec{\nabla} f\right|$ is the rate of increase of f (or slope of f) in the direction of maximum increase.

3. $\vec{\nabla} f$ is perpendicular to each level set of the function f.

4. Qualitatively, $\left|\vec{\nabla} f\right|$ is inversely proportional to the spacing between the level sets.

We can use *Maple* to graphically illustrate these properties. The level sets of a function of 2 variables may be plotted by using the `contourplot` command in the `plots` package. A 2-dimensional vector field such as the gradient of a function may be plotted by using the `fieldplot` command in the `plots` package. A 3-dimensional vector field may be plotted by using the `fieldplot3d` command in the `plots` package.

Example: 3.12. Consider the function

$$f(x, y) = \frac{\left((x-2)^2 + (y-4)^2 - 16\right)^2 - 24x - 32y + 20}{100}$$

Draw the contour plot of f using `contourplot`. Label the contours by (i) clicking in the plot on each contour to find the x and y coordinates, (ii) evaluating f at the point and (iii) using `textplot` to plot the contour values. Compute the gradient of f and plot it using `fieldplot`. Then display them all in the same plot using `display`. Finally discuss the four properties of the gradient in the context of this plot.

SOLUTION: We enter the function:

```
> f:=MF(<x,y>, (((x-2)^2 + (y-4)^2 - 16)^2 - 24*x - 32*y + 20)/100);
```

$$f := (x, y) \to \frac{1}{100}\left((x-2)^2 + (y-4)^2 - 16\right)^2 - \frac{6}{25}x - \frac{8}{25}y + \frac{1}{5}$$

We draw the contour plot and save the plot for future use:

```
> cp:=contourplot(f(x,y), x=-4..8, y=-2..10, scaling=constrained,
> contours=[-3,-2,-1,0,1,2,3,4,5]): cp;
```

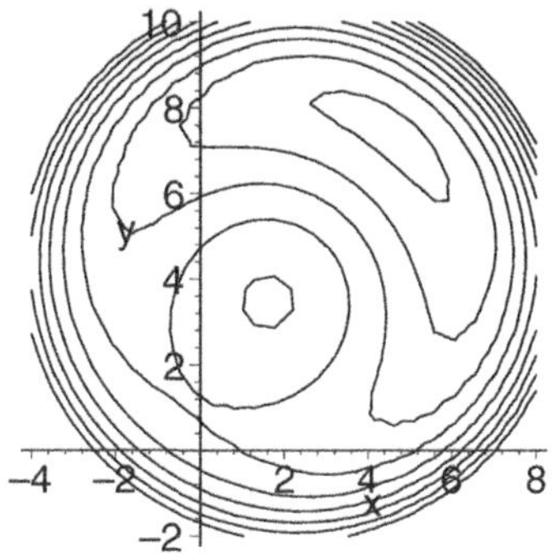

Unfortunately, *Maple* does not label the contours with the function values. So we must do it by hand. We click in the plot on each of the 10 contours, read off the coordinates and evaluate the function at that point:

```
> f(2.6,8), f(2.6,6.8), f(2.6,6), f(2.3,5.21), f(2,3.8);
> f(-1.7,2), f(-2.6,2), f(-3.4,3), f(-3.83,4), f(-3.83,6);
```

$$-2.982704000, -1.991600000, -0.9891040000, 0.0676402680, 1.051216000$$

$$-0.0034390000, 1.023056000, 2.061056000, 3.075205232, 4.034317232$$

NOTE: The actual value on each contour is the integer closest to these values. So we round to the nearest integer and use `textplot` to label each contour and `display` the text with the plot:

```
> tp:=textplot({[2.6,8,'-3'], [2.6,6.8,'-2'], [2.6,6,'-1'],
> [2.3,5.21,'0'], [2,3.8,'1'], [-1.7,2,'0'], [-2.6,2,'1'], [-3.4,3,'2'],
> [-3.83,4,'3'], [-3.83,6,'4']}, font=[TIMES,BOLD,14]):
> display({cp,tp}, scaling=constrained);
```

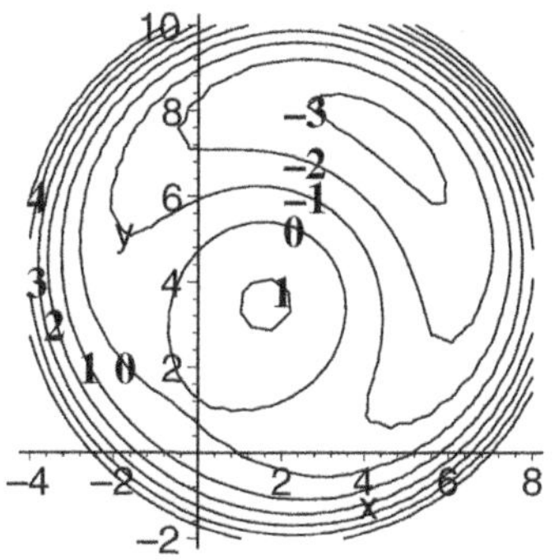

Now we compute the gradient:

```
> delf:=Grad(f,list)&@ <x,y>;
```

$$delf := [-\frac{6}{25}x^2 + \frac{1}{25}x^3 + \frac{12}{25}x - \frac{14}{25} - \frac{2}{25}y^2 + \frac{1}{25}y^2 x + \frac{16}{25}y - \frac{8}{25}y\,x,$$
$$-\frac{4}{25}x^2 + \frac{1}{25}x^2 y + \frac{16}{25}x - \frac{4}{25}y\,x - \frac{24}{25} + \frac{36}{25}y - \frac{12}{25}y^2 + \frac{1}{25}y^3]$$

and plot it, again saving the plot:

```
> fp:=fieldplot(delf, x=-3..7, y=-1..9, scaling=constrained,
> grid=[10,10], arrows=thick): fp;
```

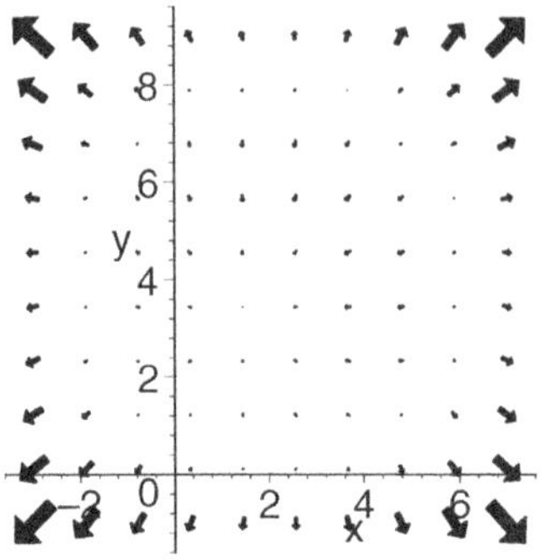

Finally, we display the level sets and the gradient field in the same plot:

```
> display({cp,tp,fp}, scaling=constrained);
```

Observe the gradient vectors are perpendicular to the level curves and point toward increasing values of the function, as required by properties 1 and 3. And the gradient vectors are longer where the contours are closer together, as predicted by properties 2 and 4. *Maple* rescales the vectors so they will fit into the plot but maintains their relative sizes. So it is not possible to observe the actual lengths of the gradient vectors.

NOTE: You may also plot the gradient of a function directly by using the `gradplot` or `gradplot3d` commands from the `plots` package.

3.2.7 Tangent Plane to a Level Surface

Recall that the plane through a point $\vec{P}$ with normal vector $\vec{N}$ is given by the equation $\vec{N}\cdot\vec{X} = \vec{N}\cdot\vec{P}$. To find the equation of the plane tangent to a level surface $F(x, y, z) = C$ of a function F at a point $\vec{P} = (a, b, c)$ we simply take the normal vector to be the gradient of F at P; i.e. $\vec{N} = \vec{\nabla}F(a, b, c)$.

Example: 3.13. In example 3.7, we found the tangent plane to the surface

$$F(x,y,z) = z^{10} + x^2y^2z^8 + x^4z^6 + y^4z^4 + x^2y^2z^2 + x^2 + 2y^2 = 8$$

at the point $(1,1,1)$ by using implicit differentiation. We now rederive it using the gradient.

SOLUTION: We define the function F, the point $P = (1,1,1)$ and the generic point $X = (x,y,z)$:

```
> F := MF([x,y,z], z^10 + x^2 * y^2 * z^8 + x^4 * z^6 + y^4 * z^4 + x^2
> * y^2 * z^2 + x^2 + 2*y^2);
```

$$F := (x, y, z) \to z^{10} + x^2\,y^2\,z^8 + x^4\,z^6 + y^4\,z^4 + x^2\,y^2\,z^2 + x^2 + 2\,y^2$$

```
> P:=<1,1,1>:  X:=<x,y,z>:
```

Then we compute the gradient and evaluate at P:

```
> delF:=Grad(F);
```

$$\begin{aligned} delF := [&(x,\, y,\, z) \to 2\,x\,y^2\,z^8 + 4\,x^3\,z^6 + 2\,x\,y^2\,z^2 + 2\,x, \\ &(x,\, y,\, z) \to 2\,x^2\,y\,z^8 + 4\,y^3\,z^4 + 2\,x^2\,y\,z^2 + 4\,y, \\ &(x,\, y,\, z) \to 10\,z^9 + 8\,x^2\,y^2\,z^7 + 6\,x^4\,z^5 + 4\,y^4\,z^3 + 2\,x^2\,y^2\,z] \end{aligned}$$

```
> N:=delF&@P;
```

$$N := [10,\ 12,\ 30]$$

Finally we construct the equation of the tangent plane:

```
> N &.  X = N &.  P;
```

$$10\,x + 12\,y + 30\,z = 52$$

Notice how much easier this was than the computation in example 3.7.

CAUTION: At the beginning and end of this chapter,we discussed the tangent plane to a surface in two different contexts. Students often confuse these two situations.

In subsection 3.2.1, we discussed the tangent plane to the graph $z = f(x,y)$ of a function of **2 variables**. In that case the tangent plane at $\big(a, b, f(a,b)\big)$ is the linear approximation:

$$z = f_{\tan}(x,y) = f(a,b) + f_x(a,b)(x-a) + f_y(a,b)(y-b).$$

In subsection 3.2.7, we discussed the tangent plane to the level surface $F(x,y,z) = C$ of a function of **3 variables**. In that case the tangent plane at $\vec{P} = (a,b,c)$ is

$$\vec{N} \cdot \vec{X} = \vec{N} \cdot \vec{P}$$

where the normal vector is the gradient at P:

$$\vec{N} = \vec{\nabla} F(a,b,c).$$

If you are given a surface as a graph $z = f(x,y)$, then you can also treat it as the level surface $F(x,y,z) = z - f(x,y) = 0$. The reverse is not always possible. In the exercises, you will be asked to find the tangent plane to several graphs by both methods.

3.3 Exercises

- Do Lab: 9.6.
- Do Projects: 10.3 and 10.4.

1. Plot the graph and the contour plot of the function $f(x, y) = y + \sqrt{x^2 + (y-2)^2}$. Discuss the shape of the contours and any local maxima and minima of the function. Notice that f is the sum of the distances from (x, y) to the point $(0, 2)$ and the line $y = 0$.

2. Use `Diff` and `value` to check each function satisfies the wave equation $\dfrac{\partial^2 y}{\partial x^2} - \dfrac{1}{c^2}\dfrac{\partial^2 y}{\partial t^2} = 0$. Make a movie of the wave for $c = 2$ by using the command like

```
> animate(subs(c=2,fa), x=-10..10, t=-5..5, view=-5..15, frames=50);
```

   Then click in the plot and click on the PLAY ARROW on the button bar. Which direction does each wave move? Explain why by looking at the function.
   $f_a = 4(x-ct)^2-(x-ct)^3$, $f_b = \cos(x+ct)$, $f_c = \exp(x-ct)$, $f_d = \mathrm{e}^{-(x-ct)^2} + 2\mathrm{e}^{-(x+ct)^2}$.

3. For each function, compute the gradient and Hessian at a general point and at the indicated point.

   (a) $f(x, y) = x^4y^3 + 4x^2y^5 - 2y^2 + 1$ at $(x, y) = (3, 1)$

   (b) $g(x, y) = \sin(x)\cos(y)$ at $(x, y) = (\frac{\pi}{3}, \frac{\pi}{2})$

   (c) $h(x, y, z) = x^4y^3 + 4x^2z^5 - 3y^2z^4$ at $(x, y) = (1, 1, 1)$

4. Find an equation of the tangent plane to the graph of each function at the indicated point. Plot the function and its tangent plane in two different colors and combine them using `display`.

   (a) $z = f(x, y) = x^2 + 2y^2$ at $(x, y) = (1, 1)$

   (b) $z = g(x, y) = xy$ at $(x, y) = (2, 3)$

5. Compute the total differential dw for the function $w = x^2y^3$ when $x = 3$, $y = 2$, $dx = .04$, and $dy = .2$.

6. The length, width and height of a box are measured to be $L = 2 \pm .03$ cm, $W = 3 \pm .02$ cm and $H = 4 \pm .01$ cm. Then the volume of the box is $V = 24 \pm \Delta V$. Use differentials to estimate the error ΔV in the computation of the volume.

7. If $u = x^4y + y^2z^3$, where $x = rse^t$, $y = rs^2e^{-t}$, and $z = r^2s\sin t$, find the value of $\dfrac{\partial u}{\partial s}$ when $r = 2$, $s = 1$ and $t = \dfrac{\pi}{2}$. Find the derivative both by forming the composition of the functions and by using the chain rule.

8. Compute the derivative of the temperature function $T(x, y, z) = z + x^2 + y^4$ along the helix $\vec{r}(t) = (4\cos(t), 4\sin(t), 3t)$ at $t = \dfrac{\pi}{4}$. Find the answer in two ways: (a) by forming the composition, and (b) by finding the derivative along the velocity.

9. Compute the directional derivative of $w = x^2y^3$ at $(x,y) = (4,-1)$ in the direction of the *unit* vector $\hat{v}$ which points in the same direction as the vector $\vec{v} = (4,3)$.

10. Find the equation of the tangent plane to the surface $x^4y^3z + xz^5 - yz^4 = 1$ at the point $(x,y,z) = (1,1,1)$ by regarding it as the level set of a function.

11. Find the equation of the tangent plane to the ellipsoid $\dfrac{x^2}{32} + \dfrac{y^2}{36} + \dfrac{z^2}{4} = 1$ at the point $(x,y,z) = (4,3,1)$ by regarding it as the level set of a function. Compare your results to example 3.5.

12. Find the equation of the tangent plane to the level surface $F(x,y,z) = z - x^2 - 2y^2 = 0$ at the point $(x,y,z) = (1,1,3)$. The answer should be the same as for exercise 4(a).

13. Find the equation of the tangent plane to the level surface $F(x,y,z) = z - xy = 0$ at the point $(x,y,z) = (2,3,6)$. The answer should be the same as for exercise 4(b).

14. Find the equation of the tangent plane to the level surface $F(x,y,z) = z - f(x,y) = 0$ at the point $(x,y,z) = \big(a,b,f(a,b)\big)$. Compare the result to the equation of the tangent plane to the graph $z = f(x,y)$ at the point $(x,y) = (a,b)$.

Chapter 4

Max-Min Problems

There are two types of max-min problems that we will discuss:

1. **Unconstrained Max-Min Problems** Here you want to find all the critical points $\vec{x}$ of a function $f(\vec{x})$ and classify each as a local maximum, a local minimum or a saddle point. The critical points are the points $\vec{x}$ where the gradient is zero, $\vec{\nabla} f(\vec{x}) = \vec{0}$. They are classified by applying the Second Derivative Test.
2. **Constrained Max-Min Problems** Here you want to find the location $\vec{x}$ and value $f(\vec{x})$ of the absolute maximum or absolute minimum of a function f where the points are constrained to lie on a level set of a function g. There are three methods of solving a constrained max-min problem:
   (a) **Eliminating a Variable** You can solve the constraint $g(\vec{x}) = C$ for one variable, substitute into the function f and reduce the problem to an unconstrained problem with one less variable.
   (b) **Parametrizing the Constraint** You can parametrize the constraint $g(\vec{x}) = C$, substitute into the function f and reduce the problem to an unconstrained problem with one less variable.
   (c) **Lagrange Multipliers** You can solve the equation $\vec{\nabla} f = \lambda \vec{\nabla} g$ along with the constraint $g(\vec{x}) = C$, for the critical points $\vec{x}$ and the Lagrange multiplier λ.

We will also consider an example with two constraints. The discussion in this chapter is all n-dimensional, although the examples are primarily 2- and 3-dimensional.

4.1 Unconstrained Max-Min Problems

4.1.1 Finding Critical Points

To find the critical points of a function $f(\vec{x})$, you need to solve the equation $\vec{\nabla} f = \vec{0}$, where $\vec{\nabla} f$ is the gradient of f.

To compute the gradient, you can use the **Gradient** command from the **VecCalc** package or its alias **Grad**. This assumes that f is arrow-defined. To set the gradient equal to zero, you can use the **equate** command from the **student** package (autoloaded by the **VecCalc** package). To solve the equations, you can try using **solve** to get exact solutions, or use **fsolve** to get decimal approximations. If *Maple* returns answers involving one or more **RootOf**'s, you can obtain all of the roots by using either the **allvalues** or **evalf** command. It may also be useful to look at a graph or a contour plot of $f(\vec{x})$ to determine the variable ranges for the **fsolve** command and to verify that you have all the solutions.

Example: 4.1. Find the location and value of all critical points of the function $f(x, y) = 3x^2y + y^3 - 3x^2 - 3y^2 + 2$.

SOLUTION: We first input the function and draw an ordinary plot and a contour plot:

```
> f:=(x,y) -> 3*x^2*y + y^3 - 3*x^2 - 3*y^2 + 2;
```

$$f := (x, y) \to 3\,x^2\,y + y^3 - 3\,x^2 - 3\,y^2 + 2$$

```
> plot3d(f(x,y), x=-3..3, y=-2..4, view=-10..10, axes=normal,
> orientation=[-15,75]);
```

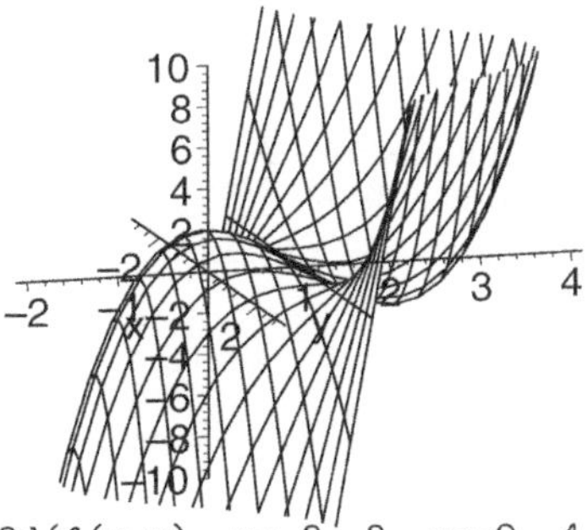

```
> contourplot3d(f(x,y), x=-3..3, y=-2..4, axes=normal, contours=[-6,
> -4, -2, -3/2, -1, -1/2, 0, 1/2, 1, 3/2, 2, 4, 6], grid=[49,49],
> orientation=[-90,0]);
```

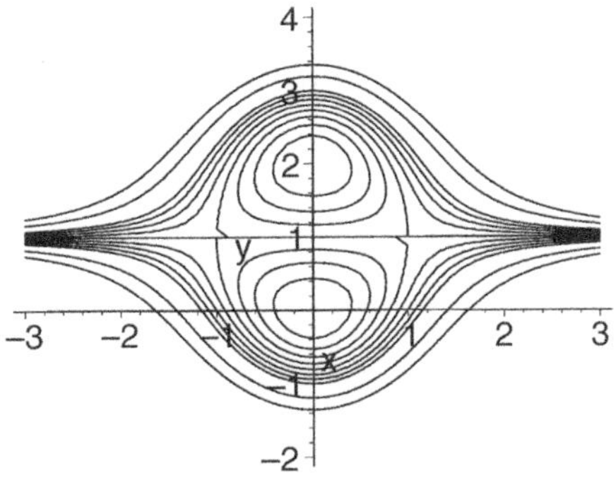

The first plot shows there are probably one local maximum and one local minimum and two saddle points. The second plot locates the local maximum near $(0, 0)$, the local minimum near $(0, 2)$ and the two saddle points near $(1, 1)$ and$(-1, 1)$.

To verify our predictions, we take the gradient of f and equate it to zero:

```
> delf:=Grad(f);
```

$$delf := [(x, y) \to 6\,y\,x - 6\,x,\ (x, y) \to 3\,x^2 + 3\,y^2 - 6\,y]$$

```
> eqs:= equate(delf &@ <x,y>, <0,0>);
```

$$eqs := \{6\,y\,x - 6\,x = 0,\ 3\,x^2 + 3\,y^2 - 6\,y = 0\}$$

We now solve the equations for the critical points:

```
> sol:=solve(eqs, {x,y});
```

$$sol := \{x = 0,\ y = 0\},\ \{x = 0,\ y = 2\},\ \{x = 1,\ y = 1\},\ \{x = -1,\ y = 1\}$$

There are four solutions, as we expected. The function values are:

```
> f1:=subs(sol[1], f(x,y)); f2:=subs(sol[2], f(x,y));
> f3:=subs(sol[3], f(x,y)); f4:=subs(sol[4], f(x,y));
```

$$f1 := 2$$
$$f2 := -2$$
$$f3 := 0$$
$$f4 := 0$$

So the local maximum is 2 at $(0, 0)$, the local minimum is -2 at $(0, 2)$ and there are two saddle points at $(1, 1)$ and $(-1, 1)$. This classification will be checked in example 4.4.

We next have a three dimensional example where the solutions are given in `RootOf` notation.

Example: 4.2. Find the location and value of all critical points of the function $F(x, y, z) = (x + y + z)\mathrm{e}^{\left(-\frac{x^2}{2} - \frac{y^2}{8} - \frac{z^2}{18}\right)}$.

SOLUTION: We first input the function. Since we cannot plot a function of 3 variables, we immediately take the gradient:

```
> F:=(x,y,z) -> (x+y+z)*exp(-x^2/2-y^2/8-z^2/18);
```

$$F := (x, y, z) \to (x + y + z)\,e^{(-1/2\,x^2 - 1/8\,y^2 - 1/18\,z^2)}$$

```
> delF:=Grad(F)&@ <x,y,z>;
```

$$delF := \left[-e^{(-\frac{x^2}{2} - \frac{y^2}{8} - \frac{z^2}{18})}(-1 + x^2 + y\,x + x\,z),\ -\frac{1}{4}e^{(-\frac{x^2}{2} - \frac{y^2}{8} - \frac{z^2}{18})}(-4 + y\,x + y^2 + y\,z),\right.$$
$$\left.-\frac{1}{9}e^{(-\frac{x^2}{2} - \frac{y^2}{8} - \frac{z^2}{18})}(-9 + x\,z + y\,z + z^2)\right]$$

Before equating the gradient to zero, we first simplify it a little bit. First notice that the exponential factor

```
> ex:=exp(-x^2/2-y^2/8-z^2/18);
```

$$ex := e^{(-\frac{x^2}{2} - \frac{y^2}{8} - \frac{z^2}{18})}$$

can be factored out of each term of the gradient. Since this quantity is always positive, we can divide each equation by this factor and not affect the validity of the equations. The resulting equations are:

```
> eqs:=equate( simplify(delF/ex), <0,0,0>);
```

$$eqs := \{1 - x^2 - y\,x - x\,z = 0,\ 1 - \frac{1}{9}x\,z - \frac{1}{9}y\,z - \frac{1}{9}z^2 = 0,\ 1 - \frac{1}{4}y\,x - \frac{1}{4}y^2 - \frac{1}{4}y\,z = 0\}$$

We now solve the equations for the critical points:

```
> sols:=solve(eqs, {x,y,z});
```

$$sols := \{x = \frac{1}{2}\mathrm{RootOf}(7\_Z^2 - 2,\ label = \_L1),\ y = 2\,\mathrm{RootOf}(7\_Z^2 - 2,\ label = \_L1),$$
$$z = \frac{9}{2}\mathrm{RootOf}(7\_Z^2 - 2,\ label = \_L1)\}$$

There is only one solution, but it involves 3 `RootOf`'s with the same arguments. The argument $7\_Z^2 - 2$ means that each `RootOf` is to be replaced by either $\sqrt{\frac{2}{7}}$ or $-\sqrt{\frac{2}{7}}$. The fact that the argument $label = \_L1$ is the same for all 3 `RootOf`'s means all 3 `RootOf`'s are to be replaced by the same root. We can force *Maple* to do these substitutions by executing

```
> sols:=allvalues(%);
```

$$sols := \{x = \frac{\sqrt{14}}{14},\ y = \frac{2\sqrt{14}}{7},\ z = \frac{9\sqrt{14}}{14}\},\ \{x = -\frac{\sqrt{14}}{14},\ y = -\frac{2\sqrt{14}}{7},\ z = -\frac{9\sqrt{14}}{14}\}$$

To get the components in alphabetical order, we execute

```
> P1:=subs(sols[1],<x|y|z>); P2:=subs(sols[2],<x|y|z>);
```

$$P1 := \left[\frac{\sqrt{14}}{14},\ \frac{2\sqrt{14}}{7},\ \frac{9\sqrt{14}}{14}\right]$$

$$P2 := \left[-\frac{\sqrt{14}}{14},\ -\frac{2\sqrt{14}}{7},\ -\frac{9\sqrt{14}}{14}\right]$$

Finally the function values are:

```
> F1:=F&@P1; F2:=F&@P2;
```

$$F1 := \sqrt{14}\,e^{(-1/2)}$$
$$F2 := -\sqrt{14}\,e^{(-1/2)}$$

So we expect that P_1 is a local maximum and P_2 is a local minimum. We will verify this in example 4.5.

In the previous two examples, we found the exact values of the critical points using the `solve` command. In the next example, the critical points cannot be found exactly. So we will use the `contourplot` and `fsolve` commands to find decimal approximations to the critical points.

Example: 4.3. Find the location and value of all critical points of the function $g(x, y) = \left((x-1)^2 + (y-2)^2 - 4\right)^2 + 3x - 4y$.

SOLUTION: We first input the function:

```
> g:=(x,y) -> ((x-1)^2+(y-2)^2-4)^2 + 3*x - 4*y;
```

$$g := (x, y) \to ((x-1)^2 + (y-2)^2 - 4)^2 + 3\,x - 4\,y$$

Next we want to plot the function. Notice that for large x and y, the function behaves qualitatively like $\left((x-1)^2 + (y-2)^2\right)^2$. So the level curves behave qualitatively like circles centered at $(1, 2)$. So we need a region centered at $(1, 2)$. Some experimentation shows that good ranges for x, y and z are $-2 \le x \le 4$, $-1 \le y \le 5$ and $-15 \le z \le 20$. Using these ranges, we draw an ordinary plot and a contour plot:

```
> plot3d(g(x,y), x=-2..4, y=-1..5, view=-15..20, style=wireframe,
> axes=boxed, orientation=[-120,75]);
```

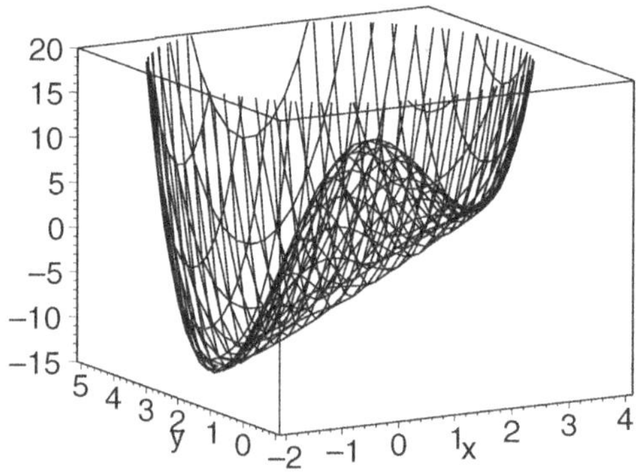


```
> contourplot3d(g(x,y), x=-2..4, y=-1..5, view=-15..20, axes=normal,
> orientation=[-90,0], grid=[49,49]);
```

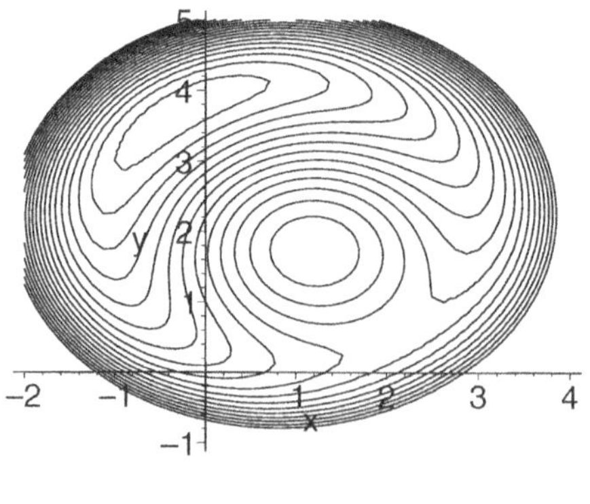

From these plots, we expect that there is a local minimum near $(-.5, 3.5)$, a local maximum near $(1, 2)$ and a saddle near $(2, .5)$.

To get better approximations for these critical points, we compute the gradient and equate it to zero:

```
> delg:=Grad(g)&@ <x,y>;
```

$$delg := [-12\,x^2 + 4\,x^3 + 12\,x - 1 - 4\,y^2 + 4\,y^2\,x + 16\,y - 16\,y\,x,\\ -8\,x^2 + 4\,x^2\,y + 16\,x - 8\,y\,x - 12 + 36\,y - 24\,y^2 + 4\,y^3]$$

```
> eqs:=equate(delg, <0,0>);
```

$$eqs := \{-12\,x^2 + 4\,x^3 + 12\,x - 1 - 4\,y^2 + 4\,y^2\,x + 16\,y - 16\,y\,x = 0,\\ -8\,x^2 + 4\,x^2\,y + 16\,x - 8\,y\,x - 12 + 36\,y - 24\,y^2 + 4\,y^3 = 0\}$$

If we solve these just using `fsolve`:

```
> sol1:=fsolve(eqs,{x,y});
```

$$sol1 := \{x = 2.092144683,\ y = 0.5438070899\}$$

we only get one solution. So we use **fsolve** with ranges to find the other two:

```
> sol2:=fsolve(eqs,{x=0..2,y=1..3});
```

$$sol2 := \{x = 1.192449828,\ y = 1.743400230\}$$

```
> sol3:=fsolve(eqs,{x=-1..0,y=3..5});
```

$$sol3 := \{x = -0.2845945104,\ y = 3.712792680\}$$

The points and function values are:

```
> Q1:=subs(sol1, <x|y>); Q2:=subs(sol2, <x|y>);
> Q3:=subs(sol3, <x|y>);
```

$$Q1 := [2.092144683,\ 0.5438070899]$$

$$Q2 := [1.192449828,\ 1.743400230]$$

$$Q3 := [-0.2845945104,\ 3.712792680]$$

```
> g1:=subs(sol1, g(x,y)); g2:=subs(sol2, g(x,y));
> g3:=subs(sol3, g(x,y));
```

$$g1 := 4.572793069$$

$$g2 := 11.79128991$$

$$g3 := -15.36408298$$

So the local minimum is -15.36 at $(-.28, 3.71)$, the local maximum is 11.79 at $(1.19, 1.74)$ and there is a saddle at $(2.09, .54)$. This will be verified in exercise 2.

4.1.2 Classifying Critical Points by the Second Derivative Test

To classify a critical point of a function f, you need to apply the Second Derivative Test. In $\mathbb{R}^2$ we have:

The Second Derivative Test in $\mathbb{R}^2$:

If (x, y) as a critical point of a function f in $\mathbb{R}^2$, then

1. (x, y) is a local minimum if $f_{xx}(x,y)f_{yy}(x,y) - f_{xy}(x,y)^2 > 0$ and $f_{xx}(x,y) > 0$.
2. (x, y) is a local maximum if $f_{xx}(x,y)f_{yy}(x,y) - f_{xy}(x,y)^2 > 0$ and $f_{xx}(x,y) < 0$.
3. (x, y) is a saddle point if $f_{xx}(x,y)f_{yy}(x,y) - f_{xy}(x,y)^2 < 0$.
4. In all other cases, the Second Derivative Test FAILS; i.e. the test cannot determine whether the critical point is a local minimum, a local maximum or a saddle point.

This may be generalized to $\mathbb{R}^n$. To do this, you first compute the Hessian matrix which is the matrix of second partial derivatives of f:

$$Hess(f) = \left(f_{x_i x_j}\right) = \begin{pmatrix} f_{x_1x_1} & f_{x_1x_2} & \cdots & f_{x_1x_n} \\ f_{x_2x_1} & f_{x_2x_2} & \cdots & f_{x_2x_n} \\ \vdots & \vdots & \ddots & \vdots \\ f_{x_nx_1} & f_{x_nx_2} & \cdots & f_{x_nx_n} \end{pmatrix}$$

Next you compute the leading principal minor determinants of the Hessian matrix. These are the determinants of the submatrices in the top left corner of sizes 1×1, 2×2, 3×3, etc:

$$D_k = \det \begin{pmatrix} f_{x_1x_1} & \cdots & f_{x_1x_k} \\ \vdots & \ddots & \vdots \\ f_{x_kx_1} & \cdots & f_{x_kx_k} \end{pmatrix} \qquad \text{for} \qquad k = 1, \ldots, n\,.$$

In particular,

$$D_1 = f_{xx}\,, \qquad D_2 = f_{xx}f_{yy} - f_{xy}^2 \qquad \text{and}$$

$$D_3 = f_{xx}f_{yy}f_{zz} + f_{xy}f_{yz}f_{zx} + f_{xz}f_{yx}f_{zy} - f_{xx}f_{yz}f_{zy} - f_{xz}f_{yy}f_{zx} - f_{xy}f_{yx}f_{zz}$$

Finally, you evaluate the leading principal minor determinants at each critical point, $\vec{x}$, and classify the point as follows:

The Second Derivative Test in $\mathbb{R}^n$:

If $\vec{x}$ as a critical point of a function f in $\mathbb{R}^n$, then

1. $\vec{x}$ is a local minimum if the determinants $D_k(\vec{x})$ are all positive.
2. $\vec{x}$ is a local maximum if the determinants $D_k(\vec{x})$ alternate signs starting with negative; i.e. $(-1)^k D_k(\vec{x}) > 0$ for $k = 1, \ldots, n$.
3. $\vec{x}$ is a saddle point if $D_n(\vec{x}) \neq 0$ and #1 and #2 above fail.
4. In all other cases, the Second Derivative Test FAILS; i.e. the test cannot determine whether the critical point is a local minimum or a local maximum.

To compute the Hessian, you can use the **Hessian** command from the **VecCalc** package or its alias **Hess**. This assumes that f is arrow-defined. The **Hess** command produces a matrix of arrow-defined functions. To display this matrix as an array of expressions, you should evaluate the Hessian at a general point using the **evalFunction** command or its alias **EF** or the operator **&@**. To compute the leading principal minor determinants, you can use the command `LeadingPrincipalMinorDeterminants` from the `VecCalc` package or its alias `LPMD`. This command expects its argument to be a matrix of expressions. So you must first evaluate the Hessian at a general point or at a critical point.

Example: 4.4. Classify the critical points of the function $f(x, y) = 3x^2y + y^3 - 3x^2 - 3y^2 + 2$.

SOLUTION: In example 4.1, we entered the function into *Maple*, found four critical points and found the function values. To classify them, we first compute and display the Hessian:

```
> Hf:=Hess(f); Hf&@ <x,y>;
```

$$Hf := \begin{bmatrix} (x, y) \to 6\,y - 6 & (x, y) \to 6\,x \\ (x, y) \to 6\,x & (x, y) \to 6\,y - 6 \end{bmatrix}$$

$$\begin{bmatrix} 6\,y - 6 & 6\,x \\ 6\,x & 6\,y - 6 \end{bmatrix}$$

Then we compute the leading principal minor determinants. For a general point we have:

```
> D1,D2:=LPMD(Hf&@ <x,y>);
```

$$\text{D1, D2} := 6\,y - 6,\ 36\,y^2 - 72\,y + 36 - 36\,x^2$$

For the first critical point $(0, 0)$, we have

```
> D1,D2:=LPMD(Hf&@ <0,0>);
```

$$\text{D1, D2} := -6,\ 36$$

Since D_2 is positive and D_1 is negative, $(0, 0)$ is a local maximum.

For the second critical point $(0, 2)$, we have

```
> D1,D2:=LPMD(Hf&@ <0,2>);
```

$$\text{D1, D2} := 6,\ 36$$

Since D_2 and D_1 are both positive, $(0, 2)$ is a local minimum.

For the third and fourth critical points $(1, 1)$ and $(-1, 1)$, we have

```
> D1,D2:=LPMD(Hf&@ <1,1>);
```

$$\text{D1, D2} := 0,\ -36$$

```
> D1,D2:=LPMD(Hf&@ <-1,1>);
```

$$\text{D1, D2} := 0,\ -36$$

At both critical points, D_2 is negative. So both $(1, 1)$ and $(-1, 1)$ are saddle points.

These results agree with our expectations from the plots in example 4.1. In particular, notice the shape of the graph and the contours near the saddle points $(1, 1)$ and $(-1, 1)$ so that you will recognize them next time.

Example: 4.5. Classify the critical points of the function $F(x, y, z) = (x+y+z)e^{\left(-\frac{x^2}{2} - \frac{y^2}{8} - \frac{z^2}{18}\right)}$.

SOLUTION: In example 4.2, we entered the function into *Maple*, found two critical points and found the function values. To classify them, we first compute the Hessian:

```
> HF:=Hess(F):
```

Then we compute the leading principal minor determinants at each critical point.

For the first critical point, $P_1 = \left(\frac{1}{\sqrt{14}}, \frac{4}{\sqrt{14}}, \frac{9}{\sqrt{14}}\right)$, we have

```
> D1,D2,D3:=LPMD(HF&@P1);
```

$$\text{D1, D2, D3} := -\frac{15}{14}\sqrt{14}\,e^{(-1/2)},\ \frac{19}{4}(e^{(-1/2)})^2,\ -\frac{7}{9}\sqrt{14}\,(e^{(-1/2)})^3$$

Since D_1 is negative, D_2 is positive and D_3 is negative, P_1 is a local maximum.

For the second critical point $P_2 = \left(-\frac{1}{\sqrt{14}}, -\frac{4}{\sqrt{14}}, -\frac{9}{\sqrt{14}}\right)$, we have

```
> D1,D2,D3:=LPMD(HF&@P2);
```

$$\text{D1, D2, D3} := \frac{15}{14}\sqrt{14}\,e^{(-1/2)},\ \frac{19}{4}(e^{(-1/2)})^2,\ \frac{7}{9}\sqrt{14}\,(e^{(-1/2)})^3$$

Since D_1, D_2 and D_3 are all positive, P_2 is a local minimum.

4.2 Constrained Max-Min Problems

Most word problems are constrained max-min problems. The foremost thing to remember when using *Maple* to solve a word problem is that *Maple* will not solve the whole problem on its own. It takes a human being to read the problem and turn the words into equations. You must identify the function to optimize and any constraint equations. Then you must choose the method of solution. Basically you should decide how you would solve the problem by hand and do the same steps using *Maple*.

One further rule of operation: If you plan to use `x`, `y`, `z` or `t` as variables in an equation, never assign a value to these variables.

4.2.1 Eliminating a Variable

One method of solving a constrained max-min problem is the "Eliminate a Variable" method. To use this method, you solve the constraint for one of the variables and substitute for that variable in the function you are extremizing. In this way, you reduce the number of variables by one. Finally, you solve the reduced unconstrained problem for the remaining variables and plug back into the constraint to find the eliminated variable.

Example: 4.6. You wish to construct an aquarium to hold 20,000 in^3 of water, with a marble base, a glass front, an aluminum back and aluminum left and right sides. There is no top. The marble costs \$.20 per in^2; the glass costs \$.12 per in^2; and the aluminum costs \$.06 per in^2. What are the dimensions which minimize the cost? (Let x be the length of the tank from left to right, y be the width from front to back, and z be the height from top to bottom.)

SOLUTION: The volume is $V = xyz$, which we enter as

```
> V:=(x,y,z) -> x*y*z;
```

$$V := (x, y, z) \to x\,y\,z$$

Then the constraint equation is

```
> constr:=V(x,y,z) = 20000;
```

$$constr := x\,y\,z = 20000$$

To find the cost, we make a table of each surface, the area of the surface and the cost per unit area:

Surface:	bottom	front	back	left	right
Area:	x y	x z	x z	y z	y z
Cost:	.20	.12	.06	.06	.06

We then multiply the cost per unit area by the area and add them up. So the total cost is

```
> C:=(x,y,z) -> .20*x*y + .12*x*z + .06*(x*z + 2*y*z);
```

$$C := (x, y, z) \to 0.20\,x\,y + 0.12\,x\,z + 0.06\,(x\,z + 2\,y\,z)$$

We now get to the first method of minimizing the cost: We solve the constraint for one variable and substitute into the cost.

```
> z0:=solve(constr,z);
```

$$z0 := \frac{20000}{x\,y}$$

```
> C2:=MF([x,y], C(x,y,z0));
```

$$C2 := (x, y) \to 0.20\,y\,x + \frac{3600.00}{y} + \frac{2400.00}{x}$$

NOTE: When we solve for z, we save the result in z0 not z so that z can still be used in other equations.

We can now minimize C_2:

```
> delC:=Grad(C2)&@ <x,y>;
```

$$delC := \left[\frac{0.2000000000\,(y\,x^2 - 12000.)}{x^2}, \frac{0.2000000000\,(x\,y^2 - 18000.)}{y^2}\right]$$

```
> eqs:=equate(delC,[0,0]);
```

$$eqs := \{\frac{0.2000000000\,(x\,y^2 - 18000.)}{y^2} = 0, \frac{0.2000000000\,(y\,x^2 - 12000.)}{x^2} = 0\}$$

```
> sol:=solve(eqs,{x,y});
```

$$sol := \{x = 20., y = 30.\},$$
$$\{x = -10.00000000 + 17.32050808\,I, y = -15.00000000 + 25.98076211\,I\},$$
$$\{x = -10.00000000 - 17.32050808\,I, y = -15.00000000 - 25.98076211\,I\}$$

Only the first solution is real. So we substitute it back into the constraint solved for z:

```
> subs(sol[1],z0);
```

$$33.33333334$$

So the dimensions are $x = 20$, $y = 30$ and $z = 33.33$. If you do not want to have decimals in your answer, then you should enter your cost function as:

```
> C:=(x,y,z)-> 20/100*x*y + 12/100*x*z + 6/100*(x*z + 2*y*z):
```

4.2.2 Parametrizing the Constraint

Another method of solving a constrained max-min problem is the "Parametrize the Constraint" method. To use this method, you parametrize the constraint set and substitute into the function to be extremized. In this way, you reduce the number of variables by one. Finally, you solve the reduced unconstrained problem for the parameters and plug back into the parametrization of the constraint to get the point.

NOTE: You should only use this method if it is easy to parametrize the constraint.

Example: 4.7. Find the point (x, y) on the ellipse $\frac{x^2}{16} + \frac{y^2}{9} = 1$ which is closest to the point $(4, 3)$.

SOLUTION: In this problem the quantity you need to minimize is the distance from the point $(4, 3)$ to the general point (x, y) on the ellipse. This distance is $d = \sqrt{(x-4)^2 + (y-3)^2}$. If the distance is a minimum, then the square of the distance is also a minimum, and vice versa. So the function we will actually minimize is

```
> f:=(x,y) -> (x-4)^2 + (y-3)^2;
```

$$f := (x, y) \to (x-4)^2 + (y-3)^2$$

The point (x, y) is constrained to lie on the ellipse. So the constraint function is

```
> g:=(x,y)->x^2/16 + y^2/9;
```

$$g := (x, y) \to \frac{1}{16}x^2 + \frac{1}{9}y^2$$

and the constraint equation is

```
> constr:=g(x,y)=1;
```

$$constr := \frac{x^2}{16} + \frac{y^2}{9} = 1$$

To solve this by the method of Eliminating a Variable, you would solve the constraint for one variable, say y:

$$y = \pm 3\sqrt{1 - \frac{x^2}{16}}$$

and substitute into the distance squared function:

$$f_+ = (x-4)^2 + (3\sqrt{1 - \frac{x^2}{16}} - 3)^2 \qquad \text{and} \qquad f_- = (x-4)^2 + (-3\sqrt{1 - \frac{x^2}{16}} - 3)^2.$$

You would then find the minima of these two functions of x. Since these are ugly functions, you would probably not choose to solve the problem by this method, (although *Maple* could handle it.) Further, for some constraints, it is not possible to solve for one variable. So you must use another method. So we turn to the method of Parametrizing the Constraint.

The ellipse may be parametrized as $x = 4\cos(\phi)$ and $y = 3\sin(\phi)$ for $0 \le \phi \le 2\pi$.

NOTE: As pointed out in example 2.1, the parameter ϕ does not measure angle like the polar coordinate θ.

This parametrization may be entered into *Maple* as

```
> x0:= 4*cos(phi):  y0:= 3*sin(phi):
```

NOTE: We store these as `x0` and `y0`, rather than as `x` and `y`, so that `x` and `y` can still be used in equations.

We then restrict the function f to the ellipse as a function of the parameter ϕ:

```
> f0:=MF( phi, f(x0,y0) );
```

$$f0 := \phi \to (4\cos(\phi) - 4)^2 + (3\sin(\phi) - 3)^2$$

We plot the function to find the number of minima and where they are.

```
> plot(f0(phi), phi=0..2*Pi);
```

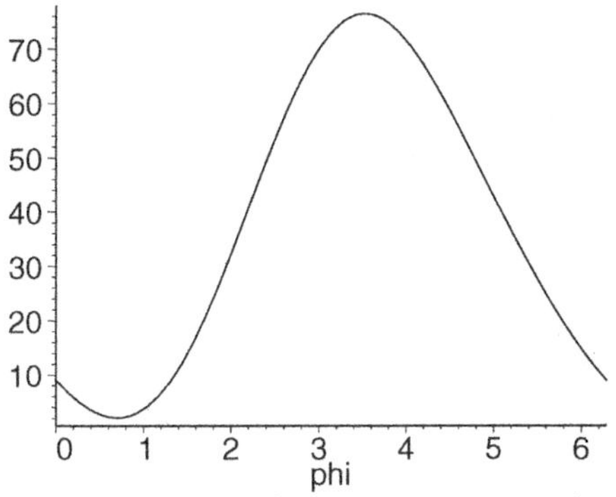

So there is one minimum near $\phi = .7$. To improve the value, we first compute the derivative:

```
> Df:=D(f0);
```

$$Df := \phi \to -8\,(4\cos(\phi) - 4)\sin(\phi) + 6\,(3\sin(\phi) - 3)\cos(\phi)$$

Then we set the derivative equal to zero and `solve`:

```
> phisol:=solve(Df(phi)=0, phi):
```

The exact solution is too complicated to print here. So we try `fsolve` with a range:

```
> phisol:=fsolve(Df(phi)=0, phi=0..2);
```

$$phisol := 0.7014935109$$

which is the minimum in the plot. Finally the x and y coordinates at the minimum are:

```
> <x|y> = evalf(subs(phi=phisol,<x0|y0>));
```

$$[x, y] = [3.055516754, 1.936077806]$$

4.2.3 Lagrange Multipliers

The final method of solving a constrained max-min problem is the method of Lagrange Multipliers. This method is based on the fact that the extremum of a function $f(\vec{x})$ along a constraint $g(\vec{x}) = C$ will occur at a point $\vec{x}$ where a level set of f is tangent to the constraint set which is itself a level set of g. Since their level sets are tangent, their normals (i.e. their gradients) are proportional, $\vec{\nabla} f = \lambda \vec{\nabla} g$. So to use this method, you solve the equations $\vec{\nabla} f = \lambda \vec{\nabla} g$ along with the constraint $g = C$ for the original variables $\vec{x}$ and the Lagrange multiplier λ. Thus the number of variables is increased by one.

Example: 4.8. Re-solve the problem of minimizing the distance from a point to an ellipse in example 4.7 using the method of Lagrange Multipliers. Also simultaneously plot: (i) the contour plot of the distance squared function, (ii) the parametric plot of the constraint ellipse and (iii) the implicit plot of the level set of the distance squared function which passes through the minimizing point. Discuss the relationship between these three pieces of the plot.

SOLUTION: Once again the square of the distance is

```
> f:=(x,y) -> (x-4)^2 + (y-3)^2:
```

the constraint function is

```
> g:=(x,y)->x^2/16 + y^2/9:
```

and the constraint equation is

```
> constr:=g(x,y)=1:
```

We compute the gradient of f and g and construct the 2 equations $\vec{\nabla} f = \lambda \vec{\nabla} g$:

```
> delf:=Grad(f) &@ <x,y>; delg:=Grad(g) &@ <x,y>;
```

$$delf := [-8 + 2\,x,\ -6 + 2\,y]$$

$$delg := \left[\frac{x}{8},\ \frac{2\,y}{9}\right]$$

```
> eqs:=equate(delf, simplify(lambda * delg));
```

$$eqs := \{-8 + 2\,x = \frac{\lambda\,x}{8},\ -6 + 2\,y = \frac{2\,\lambda\,y}{9}\}$$

To strip off the braces $\{\}$, we use the `op` command:

```
> eqs:=op(eqs);
```

$$eqs := -8 + 2\,x = \frac{\lambda\,x}{8},\ -6 + 2\,y = \frac{2\,\lambda\,y}{9}$$

Then we solve these equations and the constraint for x, y and λ:

```
> sol:=fsolve({eqs, constr}, {x,y,lambda});
```

$$sol := \{x = 3.055516754,\ y = 1.936077806,\ \lambda = -4.945720530\}$$

as found in example 4.7.

NOTE: We needed to strip the braces off `eqs` because otherwise the `fsolve` command would have an extra set of braces. The command `op` returns the operands of any expression.

So the value of f at the minimum is

```
> fsol:=subs(sol,f(x,y));
```

$$fsol := 2.023979037$$

We now turn to the plots. Recall that the parametrization of the ellipse is

```
> x0:= 4*cos(phi):  y0:= 3*sin(phi):
```

So the plot of the ellipse is

```
> ellipse:=plot([x0,y0, phi=0..2*Pi]):
```

The contour plot of f and the level curve of f with value `fsol` $= 2.024$ are

```
> fplot:=contourplot(f, -5..8, -4..5, color=black):
> fsolplot:=implicitplot(f(x,y)=fsol, x=0..6, y=0..5, thickness=4,
> color=gray):
```

Now we display them together:

```
> display({ellipse,fplot,fsolplot}, view=[-5..8, -4..5],
> scaling=constrained);
```

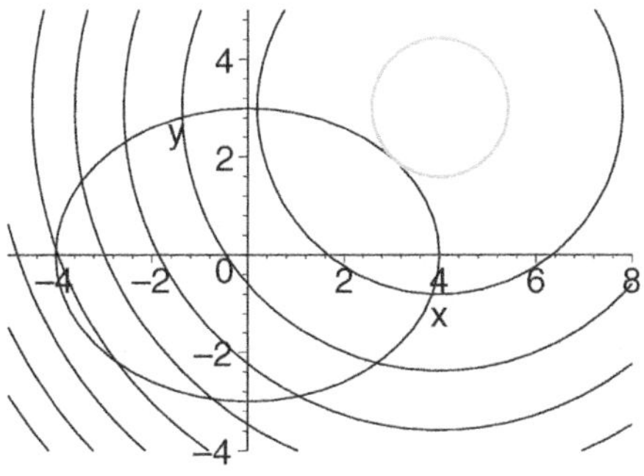

Notice that the level curves of f are concentric circles centered at $(4, 3)$. Further, the level curve of f through the minimum point, $(3.056, 1.936)$, is that contour of f which is tangent to the constraint curve (i.e. the ellipse) at the minimum point. You can also identify the maximum point as the other point of tangency between the ellipse and a level curve.

4.2.4 Two or More Constraints

In the previous examples there was one constraint relating the variables. In the following example there are two constraints. We will solve it by using Lagrange multipliers but either of the other two methods would also work.

In general, given a function $f(\vec{x})$ of n variables $\vec{x}$, suppose you want to find the location $\vec{x}$ and value $f(\vec{x})$ of the absolute maximum or absolute minimum of a function f where the points $\vec{x}$ are required to satisfy a set of k constraint equations $g_i(\vec{x}) = C_i$ for $i = 1 \ldots k$. Once again, there are three methods of solving the problem:

Eliminating k Variables If you can solve the k constraints $g_i(\vec{x}) = C_i$ for k variables, you can substitute them into the function f and reduce the problem to an unconstrained problem in $n - k$

variables. Once you solve the reduced problem, you can plug back into the constraints to find the remaining k variables.

Parametrizing the Intersection of the Constraints If you can parametrize the intersection of the k constraints $g_i(\vec{x}) = C_i$ (using $n-k$ parameters), you can substitute into the function f and reduce the problem to an unconstrained problem in the $n-k$ parameters. Once you solve the reduced problem, you can plug back into the parametrization to find the n original variables.

NOTE: This method may be the most difficult because you will need to parametrize the intersection of the k constraints in the n dimensional space.

Lagrange Multipliers To use this method, you solve the n equations $\vec{\nabla} f = \sum_{i=1}^{k} \lambda_i \vec{\nabla} g_i$ along with the k constraints $g_i(\vec{x}) = C_i$, for the n components of the critical points $\vec{x}$ and the k Lagrange multipliers λ_i. Thus there are $n+k$ equations in $n+k$ variables.

Example: 4.9. Find a point

$$\vec{p}_1 = (x_1, y_1, z_1) \quad \text{on the sphere} \quad (x-7)^2 + (y-14)^2 + (z-21)^2 = 270$$

and a point

$$\vec{p}_2 = (x_2, y_2, z_2) \quad \text{on the sphere} \quad (x-21)^2 + (y-28)^2 + (z-28)^2 = 449$$

such that the dot product $\vec{p}_1 \cdot \vec{p}_2$ is a minimum. Also find the points $\vec{p}_1$ and $\vec{p}_2$ such that the dot product $\vec{p}_1 \cdot \vec{p}_2$ is a maximum.

SOLUTION: We first enter the two points and define an abbreviation for the list of all 6 variables:

```
> p1:=[x1,y1,z1]:  p2:=[x2,y2,z2]:  ps:=op(p1),op(p2);
```

$$ps := x1,\ y1,\ z1,\ x2,\ y2,\ z2$$

NOTE: The `op` commands strip off the brackets.

We then define the constraint functions and constraint equations:

```
> g1:=MF([ps], (x1-7)^2 + (y1-14)^2 + (z1-21)^2);
```

$$g1 := (x1,\ y1,\ z1,\ x2,\ y2,\ z2) \to (x1-7)^2 + (y1-14)^2 + (z1-21)^2$$

```
> constr1:=g1(ps)=270;
```

$$constr1 := (x1-7)^2 + (y1-14)^2 + (z1-21)^2 = 270$$

```
> g2:=MF([ps], (x2-21)^2 + (y2-28)^2 + (z2-28)^2);
```

$$g2 := (x1,\ y1,\ z1,\ x2,\ y2,\ z2) \to (x2-21)^2 + (y2-28)^2 + (z2-28)^2$$

```
> constr2:=g2(ps)=449;
```

$$constr2 := (x2-21)^2 + (y2-28)^2 + (z2-28)^2 = 449$$

Notice that even though each constraint only depends on 3 coordinates, we still define it as a function of all six variables to facilitate the later computation of the gradients. The function to extremize is:

```
> f:=MF([ps],Dot(p1,p2));
```

$$f := (x1,\ y1,\ z1,\ x2,\ y2,\ z2) \to x1\ x2 + y1\ y2 + z1\ z2$$

We now compute the 3 gradients and construct the 6 equations $\vec{\nabla} f = \lambda \vec{\nabla} g_1 + \mu \vec{\nabla} g_2$:

```
> delf:=Grad(f) &@ ps; delg1:=Grad(g1) &@ ps; delg2:=Grad(g2) &@ ps;
```

$$delf := [x2,\ y2,\ z2,\ x1,\ y1,\ z1]$$
$$delg1 := [-14 + 2\,x1,\ -28 + 2\,y1,\ -42 + 2\,z1,\ 0,\ 0,\ 0]$$
$$delg2 := [0,\ 0,\ 0,\ -42 + 2\,x2,\ -56 + 2\,y2,\ -56 + 2\,z2]$$

```
> eqs:=op(equate(delf, simplify(lambda*delg1+mu*delg2)));
```

$$eqs := x1 = 2\,\mu\,(x2 - 21),\ x2 = 2\,\lambda\,(x1 - 7),\ y1 = 2\,\mu\,(y2 - 28),\ y2 = 2\,\lambda\,(y1 - 14),$$
$$z1 = 2\,\mu\,(z2 - 28),\ z2 = 2\,\lambda\,(z1 - 21)$$

We now solve the 6 equations, `eqs`, together with the 2 constraint equations, `constr1` and `constr2`, for the 8 variables, $x_1, y_1, z_1, x_2, y_2, z_2, \lambda, \mu$:

```
> sol:=solve({eqs, constr1, constr2}, {ps, lambda, mu}):
```

The two solutions are not shown because the second is very long. The first is

```
> s1:=sol[1];
```

$$s1 := \{\mu = 1,\ x1 = 14,\ x2 = 28,\ y1 = 24,\ y2 = 40,\ z1 = 32,\ z2 = 44,\ \lambda = 2\}$$

The second solution involves a `RootOf`. So we separate them by using the `allvalues` command:

```
> sol2:=evalf(allvalues(sol[2]));
```

$$sol2 := \{\mu = 0.2385928349,\ x1 = 1.062795733,\ x2 = 23.22721636,\ y1 = 4.455908036,$$
$$y2 = 37.33789154,\ z1 = 9.014178623,\ z2 = 46.89029615,\ \lambda = -1.956073541\},\{$$
$$\mu = -0.986857568,\ x1 = 15.87975236,\ x2 = 12.9543850,\ y1 = 24.63673138,$$
$$y2 = 15.5175852,\ z1 = 29.83232374,\ z2 = 12.8851931,\ \lambda = 0.72943390\},\{$$
$$\mu = -0.1890686530 - 0.09493321443\,I,\ x1 = 4.913067661 + 5.131113136\,I,$$
$$x2 = 5.181822841 - 5.626983336\,I,\ y1 = 6.074662783 + 3.185340981\,I,$$
$$y2 = 11.79192799 - 0.285537076\,I,\ z1 = 5.360521639 - 2.298873819\,I,$$
$$z2 = 19.11617427 + 10.54012401\,I,\ \lambda = -0.646710143 - 0.2419106890\,I\},\{$$
$$\mu = -0.4345735664,\ x1 = -4.822502208,\ x2 = 26.54854527,\ y1 = 3.139901963,$$
$$y2 = 24.38737579,\ z1 = 17.49475932,\ z2 = 7.871348990,\ \lambda = -1.122797222\},\{$$
$$\mu = -0.1890686530 + 0.09493321443\,I,\ x1 = 4.913067661 - 5.131113136\,I,$$
$$x2 = 5.181822841 + 5.626983336\,I,\ y1 = 6.074662783 - 3.185340981\,I,$$
$$y2 = 11.79192799 + 0.285537076\,I,\ z1 = 5.360521639 + 2.298873819\,I,$$
$$z2 = 19.11617427 - 10.54012401\,I,\ \lambda = -0.646710143 + 0.2419106890\,I\}$$

Inspecting these, we see the 2nd and 5th of these solutions involve imaginary numbers. So the remaining critical points are:

```
> s2:=sol2[1];
```

$$s2 := \{\mu = 0.2385928349,\ x1 = 1.062795733,\ x2 = 23.22721636,\ y1 = 4.455908036,$$
$$y2 = 37.33789154,\ z1 = 9.014178623,\ z2 = 46.89029615,\ \lambda = -1.956073541\}$$

```
> s3:=sol2[3];
```

$$s3 := \{\mu = -0.1890686530 - 0.09493321443\, I,\ x1 = 4.913067661 + 5.131113136\, I,\ x2 = 5.181822841 - 5.626983336\, I,\ y1 = 6.074662783 + 3.185340981\, I,\ y2 = 11.79192799 - 0.285537076\, I,\ z1 = 5.360521639 - 2.298873819\, I,\ z2 = 19.11617427 + 10.54012401\, I,\ \lambda = -0.646710143 - 0.2419106890\, I\}$$

```
> s4:=sol2[4];
```

$$s4 := \{\mu = -0.4345735664,\ x1 = -4.822502208,\ x2 = 26.54854527,\ y1 = 3.139901963,\ y2 = 24.38737579,\ z1 = 17.49475932,\ z2 = 7.871348990,\ \lambda = -1.122797222\}$$

You can check that each of these satisfy the Lagrange equations and the constraints by using commands like:

```
> subs(s1,[eqs,constr1,constr2]);
```

$$[14 = 14,\ 28 = 28,\ 24 = 24,\ 40 = 40,\ 32 = 32,\ 44 = 44,\ 270 = 270,\ 449 = 449]$$

Finally, we substitute the critical points into the function:

```
> subs(s1,f(ps));
```

$$2760$$

```
> subs(s2,f(ps));
```

$$613.7375026$$

```
> subs(s3,f(ps));
```

$$253.5759344 + 47.32442969\, I$$

```
> subs(s4,f(ps));
```

$$86.25090702$$

We see that the minimum is at s3 where the points are:

```
> 'p1' = subs(s3, p1), 'p2' = subs(s3, p2);
```

$$p1 = [4.913067661 + 5.131113136\, I,\ 6.074662783 + 3.185340981\, I,\ 5.360521639 - 2.298873819\, I],\ p2 = [5.181822841 - 5.626983336\, I,\ 11.79192799 - 0.285537076\, I,\ 19.11617427 + 10.54012401\, I]$$

and the maximum is at s1 where the points are:

```
> 'p1' = subs(s1, p1), 'p2' = subs(s1, p2);
```

$$p1 = [14,\ 24,\ 32],\ p2 = [28,\ 40,\ 44]$$

4.3 Exercises

- Do Lab: 9.7.
- Do Projects: 10.4, 10.5, 10.6, 10.7 and 10.8.

1. For the function $f(x, y) = 3x^2y + y^3 - 3x^2 - 3y^2 + 2$, find the location and value of all critical points. Then classify each critical point as a local maximum, a local minimum or a saddle point. Verify your conclusions with appropriate plots.

2. The three critical points of the function $g(x, y) = ((x-1)^2 + (y-2)^2 - 4)^2 + 3x - 4y$ were found in example 4.3. Now classify each critical point as a local maximum, a local minimum or a saddle point.

3. For the function $f(x, y) = -x^4 + 4xy - 2y^2 + 1$, find the location and value of all critical points. Then classify each critical point as a local maximum, a local minimum or a saddle point. Verify your conclusions with appropriate plots.

4. For the function $f(x, y) = -x^4 + 6xy - 2y^2 + 1$, find the location and value of all critical points. Then classify each critical point as a local maximum, a local minimum or a saddle point. Verify your conclusions with appropriate plots.

5. For the function $f(x, y) = (-14 + x^2 + y^2 - 2x + 6y)e^{(x+y)}$, find the location and value of all critical points. Then classify each critical point as a local maximum, a local minimum or a saddle point. Verify your conclusions with appropriate plots.

6. For the function $f(x, y) = ((x-1)^4 + (y-2)^4 - 4)^2 + 3x - 4y$, find the location and value of all critical points. (Use `fsolve`.) Use appropriate plots to locate the ranges for solving. Then classify each critical point as a local maximum, a local minimum or a saddle point.

7. For the function $f(x, y) = ((x-2)^4 + (y-3)^4 - 9)^2 + 3x^2 - 4y^3$, find the location and value of all critical points. (Use `solve` and `allvalues`.) Then classify each critical point as a local maximum, a local minimum or a saddle point. Verify your conclusions with appropriate plots.

8. Find the extrema of the function $f(x, y, z) = x + 2z + yz - x^2 - y^2 - z^2$.

9. Re-solve example 4.7 by Eliminating a Variable.

10. Re-solve example 4.6 by Parametrizing the Constraint.

11. Re-solve example 4.6 using Lagrange Multipliers.

12. Find the point (x, y, z) on the ellipsoid $\dfrac{x^2}{25} + \dfrac{y^2}{16} + \dfrac{z^2}{9} = 1$ which is closest to the point $(5, 4, 3)$. Use the method of Parametrizing the Constraint.

    NOTE: The ellipsoid may be parametrized by

    $$x = 5\sin(\phi)\cos(\theta), \qquad y = 4\sin(\phi)\sin(\theta), \qquad z = 3\cos(\phi)\ .$$

    Parametrized surfaces were introduced in section 1.3 and will be studied in detail in section 6.2.

13. Repeat exercise 12 but use the method of Lagrange multipliers.

14. Re-solve example 4.9 by Eliminating Two Variables.

15. Re-solve example 4.9 by Parametrizing the Two Constraints.

16. Find a point

$$\vec{p}_1 = (x_1, y_1) \quad \text{on the ellipse} \quad \frac{(x-4)^2}{16} + \frac{(y-5)^2}{25} = 1$$

and a point

$$\vec{p}_2 = (x_2, y_2) \quad \text{on the ellipse} \quad \frac{(x+3)^2}{9} + \frac{(y+4)^2}{16} = 1$$

such that the distance from $\vec{p}_1$ to $\vec{p}_2$ is a minimum. Also find the points $\vec{p}_1$ and $\vec{p}_2$ such that this distance is a maximum. Use the method of Lagrange multipliers.

17. Repeat exercise 16 but Parametrize the Two Constraints.

18. Repeat exercise 16 but Eliminate Two Variables. Plot the two ellipses to determine which half of each ellipse to use when finding the minimum and separately when finding the maximum.

19. Find a point

$$\vec{p}_1 = (x_1, y_1, z_1) \quad \text{on the ellipsoid} \quad \frac{(x-4)^2}{16} + \frac{(y-5)^2}{25} + \frac{(z-6)^2}{36} = 1$$

and a point

$$\vec{p}_2 = (x_2, y_2, z_2) \quad \text{on the ellipsoid} \quad \frac{(x+3)^2}{9} + \frac{(y+4)^2}{16} + \frac{(z+5)^2}{25} = 1$$

such that the distance from $\vec{p}_1$ to $\vec{p}_2$ is a minimum. Also find the points $\vec{p}_1$ and $\vec{p}_2$ such that this distance is a maximum. Use whichever method you prefer.

20. Find the maximum and minimum values of the function $f(x, y, z) = yz + xy$ subject to the constraints $xy = 1$ and $y^2 = 1 - z^2$.

Chapter 5

Multiple Integrals

5.1 Multiple Integrals in Rectangular Coordinates

5.1.1 Computation

To display a multiple integral such as $\int_5^6 \int_3^4 \int_1^2 x^4 y^3 z^2 \, dx \, dy \, dz$, you can use the `VecCalc` command `Multipleint` (or its alias `Muint`):

```
> Muint(x^4*y^3*z^2, x=1..2, y=3..4, z=5..6);
```

$$\int_5^6 \int_3^4 \int_1^2 x^4 \, y^3 \, z^2 \, dx \, dy \, dz$$

Then to compute its value, you use `value`:

```
> value(%);
```

$$\frac{98735}{12}$$

If you wish to bypass the display, you may use the `VecCalc` command `multipleint` (or its alias `muint`):

```
> muint(x^4*y^3*z^2, x=1..2, y=3..4, z=5..6);
```

$$\frac{98735}{12}$$

However, we recommend displaying the integral first, because you can check you have properly entered the integral. The only time you should use `muint` is in the middle of a procedure where there is no human to check the input.

Finally, if you wish to compute an integral and also see the intermediate steps in its computation, then you should use `muint` with the extra parameter "`step`". This is useful for students checking their hand computations. For example:

```
> muint(x^4*y^3*z^2, x=1..2, y=3..4, z=5..6, step):
```

$$\int_5^6 \int_3^4 \int_1^2 x^4\, y^3\, z^2\, dx\, dy\, dz$$
$$= \int_5^6 \int_3^4 \left(\frac{x^5\, y^3\, z^2}{5}\right)\Big|_{x\,=\,1..2}\, dy\, dz$$
$$= \int_5^6 \int_3^4 \frac{31\, y^3\, z^2}{5}\, dy\, dz$$
$$= \int_5^6 \left(\frac{31\, y^4\, z^2}{20}\right)\Big|_{y\,=\,3..4}\, dz$$
$$= \int_5^6 \frac{1085\, z^2}{4}\, dz$$
$$= \left(\frac{1085\, z^3}{12}\right)\Big|_{z\,=\,5..6}$$
$$= \frac{98735}{12}$$

Example: 5.1. Use a multiple integral to find the area of the general ellipse $\frac{x^2}{a^2} + \frac{y^2}{b^2} = 1$.

SOLUTION: Enter the equation into *Maple* and solve for y:

```
> ellipse:= x^2/a^2 + y^2/b^2 = 1;
```

$$ellipse := \frac{x^2}{a^2} + \frac{y^2}{b^2} = 1$$

```
> ys:=solve(ellipse,y);
```

$$ys := \frac{\sqrt{-x^2 + a^2}\, b}{a},\ -\frac{\sqrt{-x^2 + a^2}\, b}{a}$$

Then display and evaluate the integral:

```
> assume(a>0, b>0);
> Muint(1, y=ys[2]..ys[1], x=-a..a); value(%);
```

$$\int_{-a\tilde{}}^{a\tilde{}} \int_{-\frac{\sqrt{-x^2 + a\tilde{}^2}\, b\tilde{}}{a\tilde{}}}^{\frac{\sqrt{-x^2 + a\tilde{}^2}\, b\tilde{}}{a\tilde{}}} 1\, dy\, dx$$
$$b\tilde{}\ a\tilde{}\ \pi$$

Without the command `assume(a>0, b>0)`, *Maple* would not know that `a` and `b` are real and positive and so could not compute the integral. If you see tildes (˜) following the `a`'s and `b`'s (not shown above), see Appendix A to turn them off.

It is also sometimes useful to plot the region of integration. But for that you need to pick specific values for a and b, say $a = 4$ and $b = 3$.

```
> edges:=subs(a=4,b=3,{ys});
```

$$edges := \{-\frac{3\sqrt{-x^2+16}}{4}, \frac{3\sqrt{-x^2+16}}{4}\}$$

```
> plot(edges, x=-4..4, scaling=constrained);
```

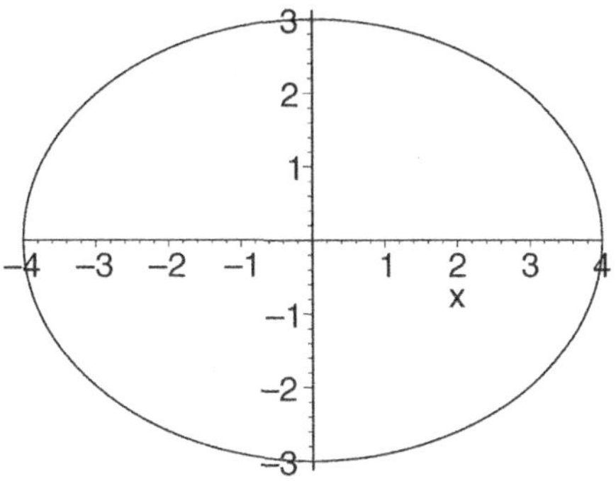

Of course, multiple integrals may be computed in any dimension:

Example: 5.2. Find the 4-dimensional volume of the 4-dimensional ball $x^2 + y^2 + z^2 + w^2 = R^2$.

SOLUTION: Display the integral and find its value:

```
> assume(R>0);
> Muint(1, w=-sqrt(R^2-x^2-y^2-z^2)..sqrt(R^2-x^2-y^2-z^2),
> z=-sqrt(R^2-x^2-y^2)..sqrt(R^2-x^2-y^2),
> y=-sqrt(R^2-x^2)..sqrt(R^2-x^2), x=-R..R);
> value(%);
```

$$\int_{-R\tilde{}}^{R\tilde{}} \int_{-\sqrt{R\tilde{}^2-x^2}}^{\sqrt{R\tilde{}^2-x^2}} \int_{-\sqrt{R\tilde{}^2-x^2-y^2}}^{\sqrt{R\tilde{}^2-x^2-y^2}} \int_{-\sqrt{R\tilde{}^2-x^2-y^2-z^2}}^{\sqrt{R\tilde{}^2-x^2-y^2-z^2}} 1\, dw\, dz\, dy\, dx$$

$$\frac{\pi^2 R\tilde{}^4}{2}$$

Once again, without the assumptions, *Maple* is unable to do the integrals.

5.1.2 Applications

§16.5 and Table C.1 in Appendix C, shows the standard applications of 2- and 3-dimensional integrals. The examples below demonstrate how to compute some of them. Other examples appear throughout the rest of this chapter and in the exercises in section 5.4.

Example: 5.3. Find the area and centroid of the region between $y = \sin(x)$ and $y = \cos(x)$ between $x = 0$ and $x = \pi/4$. Plot the region and the centroid.

SOLUTION: The area is

```
> Muint(1, y=sin(x)..cos(x), x=0..Pi/4); area:=value(%);
```

$$\int_0^{\frac{\pi}{4}} \int_{\sin(x)}^{\cos(x)} 1 \, dy \, dx$$

$$area := \sqrt{2} - 1$$

The moments about the y- and x-axes are

```
> Muint(x, y=sin(x)..cos(x), x=0..Pi/4); My:=value(%);
```

$$\int_0^{\frac{\pi}{4}} \int_{\sin(x)}^{\cos(x)} x \, dy \, dx$$

$$My := -1 + \frac{\sqrt{2}\,\pi}{4}$$

```
> Muint(y, y=sin(x)..cos(x), x=0..Pi/4); Mx:=value(%);
```

$$\int_0^{\frac{\pi}{4}} \int_{\sin(x)}^{\cos(x)} y \, dy \, dx$$

$$Mx := \frac{1}{4}$$

And the x- and y-components of the centroid are

```
> xbar, ybar:=My/area, Mx/area; evalf(%);
```

$$xbar,\ ybar := \frac{-1 + \frac{\sqrt{2}\,\pi}{4}}{\sqrt{2} - 1},\ \frac{1}{4\,(\sqrt{2} - 1)}$$

$$0.2673034979,\ 0.6035533912$$

We can see from a plot that the location of the centroid is reasonable:

```
> region:=plot({sin(x),cos(x)}, x=0..Pi/4):
> centroid:=plot([[xbar,ybar]], x=0..Pi/4, style=POINT, symbol=CIRCLE):
> display({region,centroid}, scaling=constrained);
```

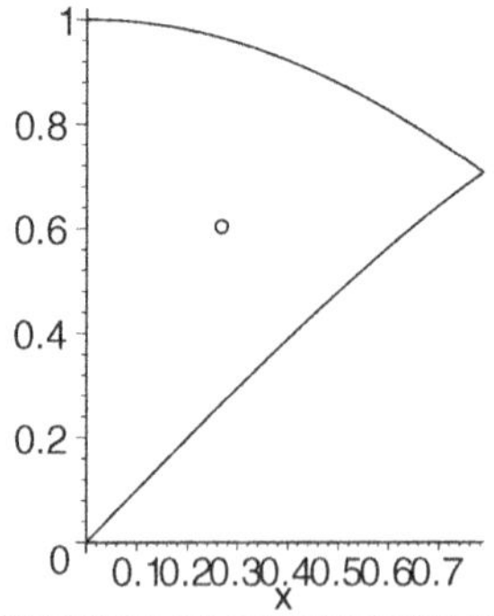

Example: 5.4. Find the mass and center of mass of the solid region in the first octant between the paraboloid $z = x^2 + y^2$ and the plane $z = 4$ if the density is given by $\rho = 1 + x + z$. Plot the solid and the center of mass.

SOLUTION: The mass is

```
> Muint(1+x+z, z=x^2+y^2..4, y=0..sqrt(4-x^2), x=0..2);
> mass:=value(%);
```

$$\int_0^2 \int_0^{\sqrt{4-x^2}} \int_{x^2+y^2}^4 1 + x + z \, dz \, dy \, dx$$

$$mass := \frac{22\,\pi}{3} + \frac{64}{15}$$

Then the moments about the yz-, xz- and xy-planes are

```
> Muint(x*(1+x+z), z=x^2+y^2..4, y=0..sqrt(4-x^2), x=0..2);
> Myz:=value(%);
```

$$\int_0^2 \int_0^{\sqrt{4-x^2}} \int_{x^2+y^2}^4 x\,(1 + x + z) \, dz \, dy \, dx$$

$$Myz := \frac{4\,\pi}{3} + \frac{576}{35}$$

```
> Muint(y*(1+x+z), z=x^2+y^2..4, y=0..sqrt(4-x^2), x=0..2);
> Mxz:=value(%);
```

$$\int_0^2 \int_0^{\sqrt{4-x^2}} \int_{x^2+y^2}^4 y\,(1 + x + z) \, dz \, dy \, dx$$

$$Mxz := \frac{2008}{105}$$

```
> Muint(z*(1+x+z), z=x^2+y^2..4, y=0..sqrt(4-x^2), x=0..2);
> Mxy:=value(%);
```

$$\int_0^2 \int_0^{\sqrt{4-x^2}} \int_{x^2+y^2}^4 z\,(1 + x + z) \, dz \, dy \, dx$$

$$Mxy := \frac{64\,\pi}{3} + \frac{256}{21}$$

And the x-, y- and z-components of the center of mass are

```
> xbar, ybar, zbar:=Myz/mass, Mxz/mass, Mxy/mass; evalf(%);
```

$$xbar,\ ybar,\ zbar := \frac{\frac{4\,\pi}{3} + \frac{576}{35}}{\frac{22\,\pi}{3} + \frac{64}{15}},\ \frac{2008}{105\,(\frac{22\,\pi}{3} + \frac{64}{15})},\ \frac{\frac{64\,\pi}{3} + \frac{256}{21}}{\frac{22\,\pi}{3} + \frac{64}{15}}$$

$$0.7561224458,\ 0.7003772406,\ 2.900973533$$

We plot the solid and the center of mass and `display` them together:

```
> solid:=plot3d({x^2+y^2,4},x=0..2,y=0..sqrt(4-x^2), axes=normal):
> cm:=spacecurve([[xbar,ybar,zbar]], style=POINT, symbol=CIRCLE):
> display({solid,cm}, scaling=constrained, orientation=[-75,75]);
```

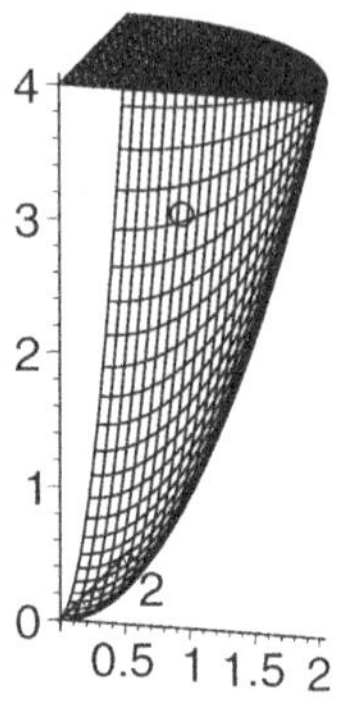

Rotate the plot and see that the center of mass is inside the solid.

Example: 5.5. Find the mass and radii of gyration of the area between the parabola $y = x^2$ and the line $y = 4$ if the density is given by $\rho = 2 + x + y$.

SOLUTION: The mass is

```
> Muint(2+x+y, y=x^2..4, x=-2..2); mass:=value(%);
```

$$\int_{-2}^{2}\int_{x^2}^{4} 2 + x + y\, dy\, dx$$

$$mass := \frac{704}{15}$$

The moments of inertia about the y- and x-axes are

```
> Muint(x^2*(2+x+y), y=x^2..4, x=-2..2); Iy:=value(%);
```

$$\int_{-2}^{2}\int_{x^2}^{4} x^2\,(2 + x + y)\, dy\, dx$$

$$Iy := \frac{4352}{105}$$

```
> Muint(y^2*(2+x+y), y=x^2..4, x=-2..2); Ix:=value(%);
```

$$\int_{-2}^{2}\int_{x^2}^{4} y^2\,(2 + x + y)\, dy\, dx$$

$$Ix := \frac{23552}{63}$$

And the x- and y-radii of gyration are

```
> xbarbar, ybarbar:=sqrt(Iy/mass), sqrt(Ix/mass); evalf(%);
```

$$xbarbar,\ ybarbar := \frac{2\sqrt{1309}}{77},\ \frac{4\sqrt{26565}}{231}$$

$$0.9397429876,\ 2.822298349$$

5.2 Multiple Integrals in Standard Curvilinear Coordinates

5.2.1 Polar Coordinates

The polar coordinate system (ρ, θ) was discussed in section 1.2.1. In polar coordinates, the Jacobian factor is r and the area differential is $dA = r\,dr\,d\theta$. So the double integral can be written as

$$\iint_R f(x, y)\,dA = \iint_R f(r, \theta)\,r\,dr\,d\theta\ .$$

Example: 5.6. Plot the cardioid $r = 1 - \sin(\theta)$ and compute the area.

SOLUTION: We first input the formula for the curve:

```
> r0:=1 - sin(theta):
```

Notice that we do not name the curve `r` so that we can still use that name for the variable in equations. Then, any of the following commands will plot the polar curve. (We only show the output from the first.) The first is the parametric form of the `plot` command:

```
> plot([r0*cos(theta), r0*sin(theta), theta=0..2*Pi] );
```

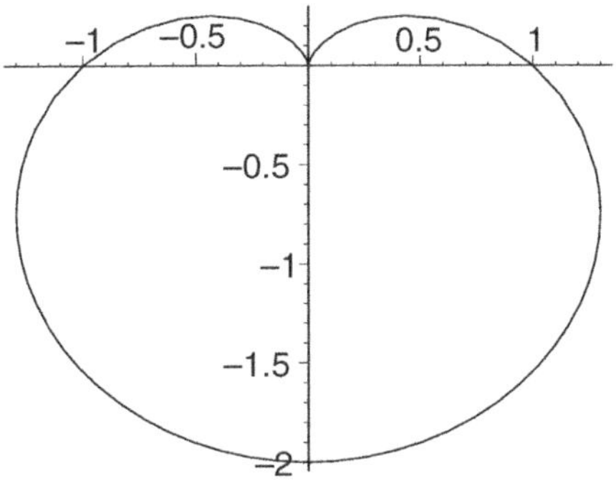

The second form is a plot command with an option which says that the coordinates are polar:

```
> plot(r0, theta=0..2*Pi, coords=polar):
```

And the third form is a specially designed command `polarplot` in the `plots` package:

```
> polarplot(r0, theta=0..2*Pi):
```

The area is

$$A = \iint 1\,dA = \int_0^{2\pi}\int_0^{1-\sin(\theta)} 1\,r\,dr\,d\theta$$

and so may be computed from

```
> Muint(r, r=0..r0, theta=0..2*Pi); A:=value(%);
```

$$\int_0^{2\,\pi}\int_0^{1-\sin(\theta)} r\,dr\,d\theta$$

$$A := \frac{3\,\pi}{2}$$

NOTE: Don't forget to include the Jacobian r in the integrand.

5.2.2 Cylindrical Coordinates

The cylindrical coordinate system (ρ, θ, z) was discussed in section 1.2.2. In cylindrical coordinates, the Jacobian factor is r and the volume differential is $dA = r\,dr\,d\theta\,dz$. So the triple integral can be written as

$$\iiint\limits_R f(x,y,z)\,dV = \iiint\limits_R f(r,\theta,z)\,r\,dr\,d\theta\,dz\ .$$

Example: 5.7. Plot the region between the paraboloids $z = x^2 + y^2$ and $z = 32 - x^2 - y^2$ but outside the cylinder $x^2 + y^2 = 4$. Then compute its volume.

SOLUTION: Rewriting the boundaries of the region in cylindrical coordinates, we find that the paraboloids are $z = r^2$ and $z = 32 - r^2$ while the cylinder is $r^2 = 4$ or $r = 2$. Before we can plot or integrate over this region, we must first understand the ranges for the coordinates. Since the paraboloids completely circle the z-axis, we have $0 \le \theta \le 2\pi$. The z coordinate is limited by the paraboloids. So it remains to find the r range. This starts at $r = 2$ and goes to the circle where the paraboloids intersect. Equating the paraboloids, we have $r^2 = 32 - r^2$ or $r = 4$.

The top and bottom paraboloids may be plotted using either of two commands. (We only show the output from the first.) The first is a `plot3d` with a parametric argument and an option specifying cylindrical coordinates:

```
> plot3d({[r, theta, r^2], [r, theta, 32-r^2]}, r=2..4,
> theta=0..2*Pi, coords=cylindrical);
```

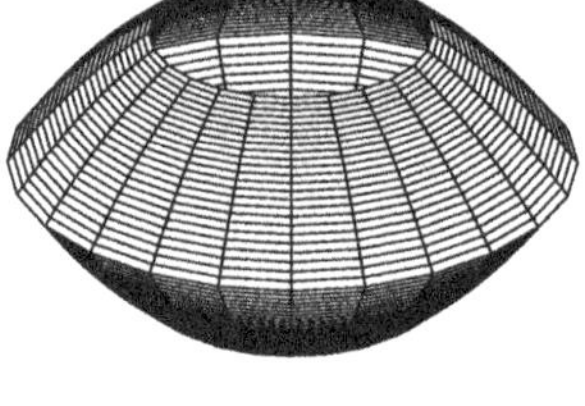

The second is the `cylinderplot` command from the `plots` package again with a parametric argument:

```
> topbot:=cylinderplot({[r, theta, r^2], [r, theta, 32-r^2]}, r=2..4,
> theta=0..2*Pi):
```

The central cylinder may also be plotted in two ways, first as a parametric `plot3d` in cylindrical coordinates:

```
> plot3d([2, theta, z], theta=0..2*Pi, z=4..28, coords=cylindrical,
> view=[-4..4, -4..4, 4..28]);
```

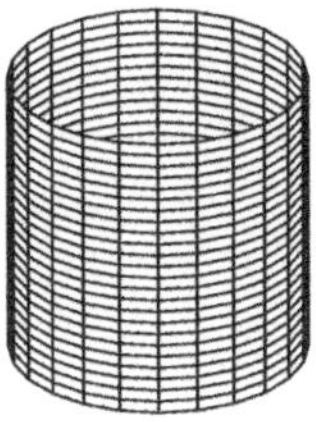

and second as a `cylinderplot`: (This is not a parametric plot because the default is that the function being plotted gives r as a function of θ and z.)

```
> inside:=cylinderplot(2, theta=0..2*Pi, z=4..28):
```

The top, bottom and inside surfaces may be put together using the `display` command:

```
> display({topbot,inside}, orientation=[45,45]);
```

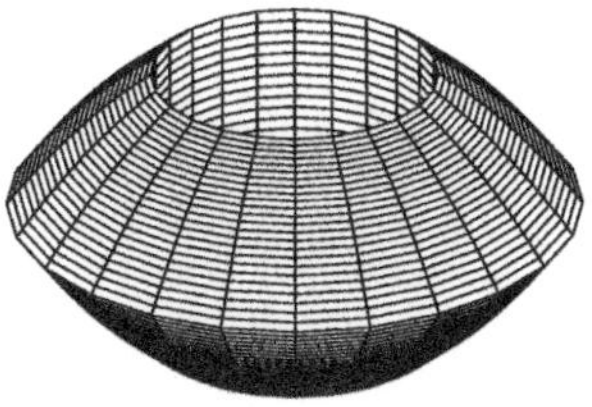

The volume is

$$V = \iiint 1\,dV = \int_0^{2\pi} \int_2^4 \int_{r^2}^{32-r^2} r\,dz\,dr\,d\theta$$

and so may be computed from

```
> Muint(r, z=r^2..32-r^2, r=2..4, theta=0..2*Pi); V:=value(%);
```

$$\int_0^{2\,\pi} \int_2^4 \int_{r^2}^{32-r^2} r\,dz\,dr\,d\theta$$

$$V := 144\,\pi$$

NOTE: You must remember to include the Jacobian r in the integrand.

5.2.3 Spherical Coordinates

The spherical coordinate system (ρ, θ, ϕ) was discussed in section 1.2.2. In spherical coordinates, the Jacobian factor is $\rho^2 \sin(\phi)$ and the volume differential is $dV = \rho^2 \sin(\phi)\,d\rho\,d\theta\,d\phi$. So the triple integral can be written as

$$\iiint\limits_R f(x,y,z)\,dV = \iiint\limits_R f(\rho,\theta,\phi)\,\rho^2 \sin(\phi)\,d\rho\,d\theta\,d\phi\ .$$

Example: 5.8. One cell of the spherical coordinate system is shown in the Figure. The coordinate ranges are $\rho_1 \le \rho \le \rho_2$, $\theta_1 \le \theta \le \theta_2$ and $\phi_1 \le \phi \le \phi_2$.

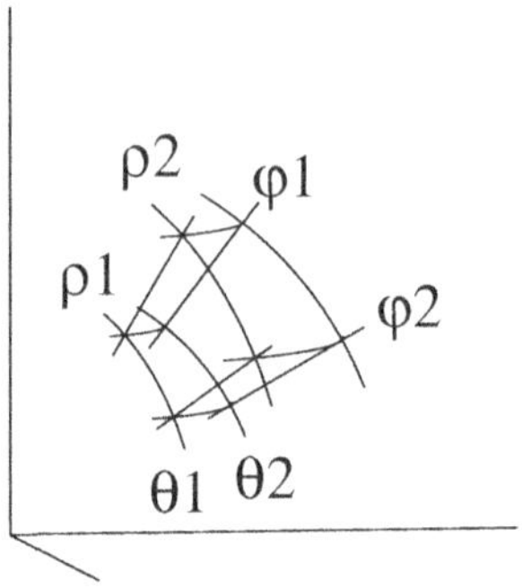

The "coordinate center" of the cell is at $(\rho_0, \theta_0, \phi_0) = \left(\dfrac{\rho_1+\rho_2}{2}, \dfrac{\theta_1+\theta_2}{2}, \dfrac{\phi_1+\phi_2}{2}\right)$ and the "coordinate dimensions" of the cell are $\Delta\rho = \rho_2 - \rho_1$, $\Delta\theta = \theta_2 - \theta_1$ and $\Delta\phi = \phi_2 - \phi_1$. In terms of these, show that the volume of the spherical cell is

$$\Delta V = \left[(\rho_0)^2 + \frac{(\Delta\rho)^2}{12}\right] \sin(\phi_0)\, \Delta\rho\, \Delta\theta\, 2\sin\left(\frac{\Delta\phi}{2}\right) .$$

Then compute $\displaystyle\lim_{(\Delta\rho,\Delta\theta,\Delta\phi)=(0,0,0)} \frac{\Delta V}{\Delta\rho\Delta\theta\Delta\phi}$ to "derive" the spherical Jacobian $J = (\rho_0)^2 \sin(\phi_0)$.

SOLUTION: The volume integral is

$$V = \iiint 1\, dV = \int_{\phi_1}^{\phi_2} \int_{\theta_1}^{\theta_2} \int_{\rho_1}^{\rho_2} 1\, \rho^2 \sin(\phi)\, d\rho\, d\theta\, d\phi$$

NOTE: Remember to include the Jacobian $\rho^2 \sin(\phi)$ in the integrand.

It may be computed from

```
> Muint( rho^2*sin(phi), rho=rho1..rho2, theta=theta1..theta2,
> phi=phi1..phi2 ); Delta_V:=value(%);
```

$$\int_{\phi 1}^{\phi 2} \int_{\theta 1}^{\theta 2} \int_{\rho 1}^{\rho 2} \rho^2 \sin(\phi)\, d\rho\, d\theta\, d\phi$$

$$Delta\_V := \frac{1}{3}\cos(\phi 1)\,\rho 2^3\,\theta 2 - \frac{1}{3}\cos(\phi 1)\,\rho 2^3\,\theta 1 - \frac{1}{3}\cos(\phi 1)\,\rho 1^3\,\theta 2 + \frac{1}{3}\cos(\phi 1)\,\rho 1^3\,\theta 1$$
$$- \frac{1}{3}\cos(\phi 2)\,\rho 2^3\,\theta 2 + \frac{1}{3}\cos(\phi 2)\,\rho 2^3\,\theta 1 + \frac{1}{3}\cos(\phi 2)\,\rho 1^3\,\theta 2 - \frac{1}{3}\cos(\phi 2)\,\rho 1^3\,\theta 1$$

To simplify this we first change variables to the average values and the widths. The equations are

```
> eqs:=[rho1=rho0-Drho/2, rho2=rho0+Drho/2, theta1=theta0-Dtheta/2,
> theta2=theta0+Dtheta/2, phi1=phi0-Dphi/2, phi2=phi0+Dphi/2]:
```

After some experimentation, it is found that the best simplification is obtained from

```
> Delta_V2:=factor( expand( subs( eqs, Delta_V )));
```

$$Delta\_V2 := \frac{1}{6}\sin(\phi 0)\sin(\frac{Dphi}{2})\, Drho\, Dtheta\,(12\,\rho 0^2 + Drho^2)$$

Finally we compute the limit:

```
> Limit(Limit(Limit('Delta_V2'/(Drho*Dtheta*Dphi), Dphi=0), Dtheta=0),
> Drho=0); J:=value(%);
```

$$\lim_{Drho\to 0}\ \lim_{Dtheta\to 0}\ \lim_{Dphi\to 0} \frac{Delta\_V2}{Drho\ Dtheta\ Dphi}$$

$$J := \sin(\phi 0)\,\rho 0^2$$

which is the spherical Jacobian.

NOTE: This "derivation" is circular since we used the Jacobian in writing the integral. A rigorous derivation is given in example 5.16.

5.2.4 Applications

Table C.1 in Appendix C, shows the standard applications of 2- and 3-dimensional integrals. The examples in subsection 5.1.2 showed how to compute these quantities in rectangular coordinates. In this section, the quantities are computed in polar, cylindrical and spherical coordinates. More examples appear in the exercises in section 5.4.

Example: 5.9. Find the centroid of the cardioid $r = 1 - \sin(\theta)$ expressed in polar coordinates.

SOLUTION: We first input the curve and the polar formulas for x and y:

```
> r0:=1 - sin(theta):
> x0:=r * cos(theta):  y0:=r * sin(theta):
```

In example 5.6, we found the area of the cardioid, $A = \int_0^{2\pi}\int_0^{1-\sin(\theta)} r\,dr\,d\theta$, to be:

```
> Muint(r, r=0..r0, theta=0..2*Pi):  A:=value(%);
```

$$A := \frac{3\pi}{2}$$

Then the first moments about the y- and x-axes are

$$M_y = \int_0^{2\pi}\int_0^{1-\sin(\theta)} x\,r\,dr\,d\theta \quad\text{and}\quad M_x = \int_0^{2\pi}\int_0^{1-\sin(\theta)} y\,r\,dr\,d\theta$$

except that x and y must be expressed in polar coordinates. Thus we compute:

```
> Muint(x0*r, r=0..r0, theta=0..2*Pi); My:=value(%);
```

$$\int_0^{2\pi}\int_0^{1-\sin(\theta)} r^2\cos(\theta)\,dr\,d\theta$$

$$My := 0$$

```
> Muint(y0*r, r=0..r0, theta=0..2*Pi); Mx:=value(%);
```

$$\int_0^{2\pi}\int_0^{1-\sin(\theta)} r^2\sin(\theta)\,dr\,d\theta$$

$$Mx := -\frac{5\pi}{4}$$

Then the x- and y-components of the centroid are

```
> xbar, ybar:=My/A, Mx/A;
```

$$xbar,\ ybar := 0,\ \frac{-5}{6}$$

Finally, we convert to polar coordinates:

```
> cm:=r2p([xbar,ybar]);
```

$$cm := [\frac{5}{6}, -\frac{\pi}{2}]$$

Thus $\overline{r} = \frac{5}{6}$ and $\overline{\theta} = -\frac{\pi}{2}$. As should be expected from the plot in example 5.6, the centroid is along the negative y-axis.

CAUTION: It is tempting to try to compute the r-component of the centroid directly as

$$\overline{r} = \frac{1}{A}\int_0^{2\pi}\int_0^{1-\sin(\theta)} r\,r\,dr\,d\theta$$

by putting an r into the moment integral instead of an x or y. This is ABSOLUTELY WRONG! It leads to the incorrect result:

```
> 1/A*Muint(r^2, r=0..r0, theta=0..2*Pi); value(%);
```

$$\frac{2}{3}\left(\frac{1}{\pi}\int_0^{2\pi}\int_0^{1-\sin(\theta)} r^2\,dr\,d\theta\right)$$

$$\frac{10}{9}$$

Example: 5.10. Find the mass and moment of inertia about the z-axis of the solid between the paraboloids $z = x^2 + y^2$ and $z = 8 - x^2 - y^2$ with density $\rho = 1 + x^2 + y^2$.

SOLUTION: We first input the paraboloids and density but in cylindrical coordinates:

```
> z1:=r^2: z2:=8-r^2: rho:=1+r^2:
```

Equating the two paraboloids, $r^2 = 8 - r^2$, we find they intersect at $r = 2$. So the mass is given by

```
> Muint(rho*r, z=z1..z2, r=0..2, theta=0..2*Pi); M:=value(%);
```

$$\int_0^{2\pi}\int_0^{2}\int_{r^2}^{8-r^2} (1+r^2)\,r\,dz\,dr\,d\theta$$

$$M := \frac{112\pi}{3}$$

Notice that you need to get the order of integration correct since the z limits depend on r. Finally, the moment of inertia about the z-axis is

```
> Muint(r^2*rho*r, z=z1..z2, r=0..2, theta=0..2*Pi); Iz:=value(%);
```

$$\int_0^{2\pi}\int_0^2\int_{r^2}^{8-r^2} r^3\,(1+r^2)\,dz\,dr\,d\theta$$

$$Iz := 64\,\pi$$

Example: 5.11. Find the mass and center of mass of a hemisphere of radius a if its density is proportional to the distance from the center of the base.

NOTE: To avoid confusion between the density and the spherical radial coordinate ρ, you should call the density δ. You will also need to clear out the variable `rho`.

```
> rho:='rho':
```

SOLUTION: We take the base to lie in the xy-plane with the center at the origin. Then the distance from the center of the base is the spherical coordinate ρ and the density is

```
> delta:=K * rho:
```

where K is a proportionality constant. Then the mass is

```
> Muint( delta * rho^2*sin(phi), rho=0..a, theta=0..2*Pi, phi=0..Pi/2);
> M:=value(%);
```

$$\int_0^{\frac{\pi}{2}}\int_0^{2\pi}\int_0^{a} K\,\rho^3\sin(\phi)\,d\rho\,d\theta\,d\phi$$

$$M := \frac{K\,a^4\,\pi}{2}$$

(Don't forget the Jacobian $\rho^2\sin(\phi)$.) By symmetry, the center of mass must be on the z-axis. So it remains to compute the z-component of the center of mass. In spherical coordinates, the z-coordinate is

```
> z0:=rho*cos(phi):
```

Then the moment away from the xy-plane is

```
> Muint( z0 * delta * rho^2*sin(phi), rho=0..a, theta=0..2*Pi,
> phi=0..Pi/2); Mxy:=value(%);
```

$$\int_0^{\frac{\pi}{2}}\int_0^{2\pi}\int_0^{a} \rho^4\cos(\phi)\,K\sin(\phi)\,d\rho\,d\theta\,d\phi$$

$$Mxy := \frac{K\,\pi\,a^5}{5}$$

and the z-component of center of mass is

```
> zbar:=Mxy/M;
```

$$zbar := \frac{2\,a}{5}$$

Notice that this is reasonable because the center of mass is inside the hemisphere.

5.3 Multiple Integrals in General Curvilinear Coordinates

5.3.1 General Curvilinear Coordinates

A curvilinear coordinate system in $\mathbb{R}^n$ is a list of n functions of n variables giving the n rectangular coordinates as functions of the n curvilinear coordinates. In general,

$$\begin{aligned}(x_1,x_2,\ldots,x_n) &= \vec{R}(u_1,u_2,\ldots,u_n)\\ &= \big(x_1(u_1,u_2,\ldots,u_n), x_2(u_1,u_2,\ldots,u_n),\ldots,x_n(u_1,u_2,\ldots,u_n)\big)\end{aligned}$$

or more briefly,

$$\vec{x} = \vec{R}(\vec{u}) \ .$$

Of course, the variable names could change. A function of this type is also called a vector function of several variables. In particular, a general curvilinear coordinate system in $\mathbb{R}^2$ has the form

$$(x,y) = \vec{R}(u,v) = \big(x(u,v), y(u,v)\big) \ .$$

and a general curvilinear coordinate system in $\mathbb{R}^3$ has the form

$$(x,y,z) = \vec{R}(u,v,w) = \big(x(u,v,w), y(u,v,w), z(u,v,w)\big) \ .$$

Throughout this section, we will look at two examples, one in $\mathbb{R}^2$ and one in $\mathbb{R}^3$:

- the 2-dimensional bipolar coordinate system given by

$$(x,y) = \vec{R}(u,v) = \left(\frac{\sinh v}{\cosh v - \cos u}, \frac{\sin u}{\cosh v - \cos u}\right)$$

- and the 3-dimensional paraboloidal coordinate system given by

$$(x,y,z) = \vec{R}(u,v,\theta) = \left(uv\cos\theta, uv\sin\theta, \frac{u^2-v^2}{2}\right) \ .$$

Maple already knows about a large number of curvilinear coordinate systems (including bipolar and paraboloidal). A complete list may be found by looking at the help page:

```
> ?coords
```

Additional coordinate systems may be added using the command `addcoords` from the `plots` package, but that is beyond this book. For more information see

```
> ?addcoords
```

For the purposes of this book, a curvilinear coordinate system may be entered into *Maple* using the `VecCalc` command `MakeFunction` or its alias `MF`. The first argument is the vector of curvilinear coordinates, and the second argument is the vector of expressions for the rectangular coordinates.

Example: 5.12. Enter the (a) bipolar and (b) paraboloidal coordinate systems into *Maple*.

SOLUTION: (a) The bipolar coordinate system is

```
> R2:=MF(<u|v>, <sinh(v)/(cosh(v)-cos(u))|
> sin(u)/(cosh(v)-cos(u))>);
```

$$R2 := \left[(u, v) \to -\frac{\sinh(v)}{-\cosh(v)+\cos(u)}, (u, v) \to -\frac{\sin(u)}{-\cosh(v)+\cos(u)}\right]$$

(b) The paraboloidal coordinate system is

```
> R3:=MF(<u|v|theta>, <u*v*cos(theta)| u*v*sin(theta)| (u^2 -
> v^2)/2>);
```

$$R3 := \left[(u, v, \theta) \to u\,v\cos(\theta), (u, v, \theta) \to u\,v\sin(\theta), (u, v, \theta) \to \frac{1}{2}u^2 - \frac{1}{2}v^2\right]$$

A coordinate curve is the curve obtained by allowing one curvilinear coordinate to vary while the other coordinates are held fixed. If you draw several coordinate curves for each coordinate, you obtain a coordinate grid for the curvilinear coordinate system. For the coordinate systems which are already known to *Maple* the command **coordplot** from the **plots** package will plot a 2 dimensional coordinate grid and the command **coordplot3d** will plot an abbreviated 3 dimensional coordinate grid (also showing the coordinate surfaces).

Example: 5.13. Plot a coordinate grid for the (a) bipolar and (b) paraboloidal coordinate systems.

SOLUTION: (a) A coordinate grid for bipolar coordinates is

```
> coordplot(bipolar);
```

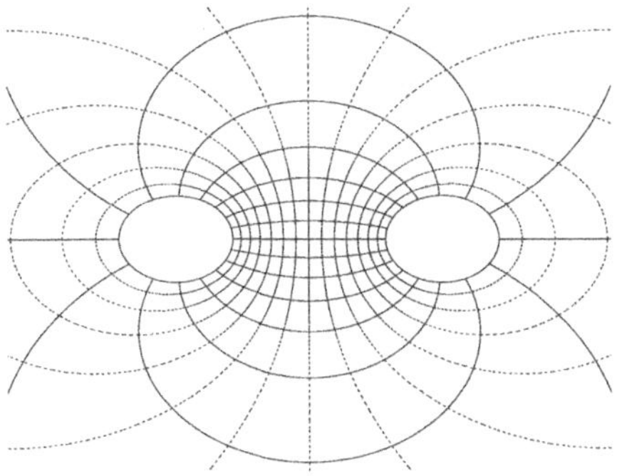

(b) A coordinate grid for paraboloidal coordinates is

```
> coordplot3d(paraboloidal);
```

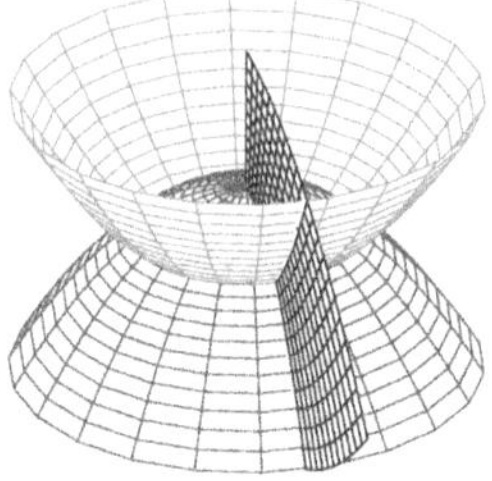

A coordinate tangent vector to a coordinate curve is obtained by differentiating with respect to the parameter on that curve, i.e. the curvilinear coordinate which is varying. Since the remaining variables are held fixed, these are partial derivatives. Thus, in a general 3-dimensional coordinate system, the three coordinate tangent vectors are

$$R_u = \frac{\partial R}{\partial u}, \quad R_v = \frac{\partial R}{\partial v}, \quad R_w = \frac{\partial R}{\partial w}$$

In *Maple*, if the coordinate system has been defined using `MF`, these may be computed using `D`.

Example: 5.14. Compute the coordinate tangent vectors for the (a) bipolar and (b) paraboloidal coordinate systems.

SOLUTION: (a) The coordinate tangent vectors for the bipolar coordinate system are:

```
> R2u:=D[1](R2); R2v:=D[2](R2);
```

$$R2u := \left[(u, v) \to -\frac{\sinh(v)\sin(u)}{(-\cosh(v)+\cos(u))^2},\right.$$
$$\left.(u, v) \to -\frac{\cos(u)}{-\cosh(v)+\cos(u)} - \frac{\sin(u)^2}{(-\cosh(v)+\cos(u))^2}\right]$$

$$R2v := \left[(u, v) \to -\frac{\cosh(v)}{-\cosh(v)+\cos(u)} - \frac{\sinh(v)^2}{(-\cosh(v)+\cos(u))^2},\right.$$
$$\left.(u, v) \to -\frac{\sinh(v)\sin(u)}{(-\cosh(v)+\cos(u))^2}\right]$$

(b) The coordinate tangent vectors for the paraboloidal coordinate system are:

```
> R3u:=D[1](R3); R3v:=D[2](R3); R3theta:=D[3](R3);
```

$$R3u := [(u, v, \theta) \to v\cos(\theta), (u, v, \theta) \to v\sin(\theta), (u, v, \theta) \to u]$$
$$R3v := [(u, v, \theta) \to u\cos(\theta), (u, v, \theta) \to u\sin(\theta), (u, v, \theta) \to -v]$$
$$R3theta := [(u, v, \theta) \to -u\,v\sin(\theta), (u, v, \theta) \to u\,v\cos(\theta), 0]$$

The Jacobian matrix of a 2-dimensional curvilinear coordinate system is the matrix whose columns are the coordinate tangent vectors $\vec{R}_u$ and $\vec{R}_v$:

$$\frac{d(x,y)}{d(u,v)} = \begin{pmatrix} \frac{\partial x}{\partial u} & \frac{\partial x}{\partial v} \\ \frac{\partial y}{\partial u} & \frac{\partial y}{\partial v} \end{pmatrix}$$

Its determinant is the Jacobian determinant:

$$\frac{\partial(x,y)}{\partial(u,v)} = \left|\frac{d(x,y)}{d(u,v)}\right| = \det\begin{pmatrix} \frac{\partial x}{\partial u} & \frac{\partial x}{\partial v} \\ \frac{\partial y}{\partial u} & \frac{\partial y}{\partial v} \end{pmatrix}$$

And the absolute value of the determinant is the Jacobian factor:

$$J(u,v) = \left|\frac{\partial(x,y)}{\partial(u,v)}\right| = \left\|\frac{d(x,y)}{d(u,v)}\right\| = \left|\det\begin{pmatrix} \frac{\partial x}{\partial u} & \frac{\partial x}{\partial v} \\ \frac{\partial y}{\partial u} & \frac{\partial y}{\partial v} \end{pmatrix}\right|$$

Informally, any of the three may be called the Jacobian. Similar definitions hold in 3 and higher dimensions.

In *Maple*, the Jacobian matrix is computed using the `VecCalc` command `JacobianMatrix` or its alias `Jac`. The Jacobian determinant is computed using the `VecCalc` command `JacobianDeterminant` or its alias `JDet` and the Jacobian factor is computed by taking the absolute value of the Jacobian determinant or more often by simply changing the sign of the determinant when necessary.

Example: 5.15. Compute the Jacobian matrix, determinant and factor for the (a) bipolar and (b) paraboloidal coordinate systems.

SOLUTION: (a) For the bipolar coordinate system these are:

```
> JM:=Jac(R2) &@ <u,v>;
```

$$JM := \begin{bmatrix} -\frac{\sinh(v)\sin(u)}{\%1} & -\frac{\cosh(v)\cos(u)-1}{\%1} \\ \frac{\cosh(v)\cos(u)-1}{\%1} & -\frac{\sinh(v)\sin(u)}{\%1} \end{bmatrix}$$

$$\%1 := \cosh(v)^2 - 2\cosh(v)\cos(u) + \cos(u)^2$$

```
> JD:=factor(JDet(R2) &@ <u,v>);
```

$$JD := \frac{1}{(\cosh(v)-\cos(u))^2}$$

Notice that the Jacobian determinant is everywhere positive (except at the origin). So the Jacobian factor is:

```
> J:=JD;
```

$$J := \frac{1}{(\cosh(v)-\cos(u))^2}$$

(b) For the paraboloidal coordinate system we have:

```
> JM:=Jac(R3) &@ <u,v,w>;
```

$$JM := \begin{bmatrix} v\cos(w) & u\cos(w) & -u\,v\sin(w) \\ v\sin(w) & u\sin(w) & u\,v\cos(w) \\ u & -v & 0 \end{bmatrix}$$

```
> JD:=factor(JDet(R3) &@ <u,v,w>);
```

$$JD := v\,u\,(v^2+u^2)$$

Notice that the Jacobian determinant is positive in the 1st and 3rd quadrants of the uv-plane and negative in the 2nd and 4th quadrants. So the Jacobian factor is:

```
> J:=abs(JD);
```

$$J := \left|v\,u\,(v^2+u^2)\right|$$

5.3.2 Multiple Integrals

The differential of area is the product of the Jacobian factor and the differentials of the curvilinear coordinates, $dA = J(u,v)\,du\,dv$. So an integral over a region R in $\mathbb{R}^2$ has the form

$$\iint_R f\,dA = \iint_R f(x,y)\,dx\,dy = \iint_R f(u,v)\,J(u,v)\,du\,dv.$$

Similarly, the differential of volume is $dV = J(u,v,w)\,du\,dv\,dw$ and an integral over a region R in $\mathbb{R}^3$ has the form

$$\iiint_R f\,dV = \iiint_R f(x,y,z)\,dx\,dy\,dz = \iiint_R f(u,v,w)\,J(u,v,w)\,du\,dv\,dw.$$

Further, the differential of n-dimensional volume is $dV = J(u_1,u_2,\ldots,u_n)\,du_1\,du_2\cdots du_n$ and an integral over a region R in $\mathbb{R}^n$ has the form

$$\int\cdots\int_R f\,dV = \int\cdots\int_R f(x_1,x_2,\ldots x_n)\,dx_1\,dx_2\cdots dx_n \tag{5.1}$$

$$= \int\cdots\int_R f(u_1,u_2,\ldots,u_n)J(u_1,u_2,\ldots,u_n)\,du_1\,du_2\cdots du_n\ . \tag{5.2}$$

Example: 5.16. For spherical coordinates,

describe the coordinate curves and compute the coordinate tangent vectors, the Jacobian matrix, the Jacobian determinant, the Jacobian factor and the volume element.

SOLUTION: The spherical coordinate system is given by

```
> R:=MF(<rho, theta, phi>, <rho*sin(phi)*cos(theta),
> rho*sin(phi)*sin(theta), rho*cos(phi)>);
```

$$R := \begin{bmatrix} (\rho,\, \theta,\, \phi) \to \rho \sin(\phi) \cos(\theta) \\ (\rho,\, \theta,\, \phi) \to \rho \sin(\phi) \sin(\theta) \\ (\rho,\, \theta,\, \phi) \to \rho \cos(\phi) \end{bmatrix}$$

The ρ-lines are the radial lines, the θ-lines are the lines of latitude, and the ϕ-lines are the lines of longitude. The coordinate tangent vectors are

```
> Rr:=D[1](R); Rtheta:=D[2](R); Rphi:=D[3](R);
```

$$Rr := \begin{bmatrix} (\rho,\, \theta,\, \phi) \to \sin(\phi) \cos(\theta) \\ (\rho,\, \theta,\, \phi) \to \sin(\phi) \sin(\theta) \\ (\rho,\, \theta,\, \phi) \to \cos(\phi) \end{bmatrix}$$

$$Rtheta := \begin{bmatrix} (\rho,\, \theta,\, \phi) \to -\rho \sin(\phi) \sin(\theta) \\ (\rho,\, \theta,\, \phi) \to \rho \sin(\phi) \cos(\theta) \\ 0 \end{bmatrix}$$

$$Rphi := \begin{bmatrix} (\rho,\, \theta,\, \phi) \to \rho \cos(\phi) \cos(\theta) \\ (\rho,\, \theta,\, \phi) \to \rho \cos(\phi) \sin(\theta) \\ (\rho,\, \theta,\, \phi) \to -\rho \sin(\phi) \end{bmatrix}$$

The Jacobian matrix and the Jacobian determinant are

```
> Jac(R) &@ <rho, theta, phi>; JDet(R) &@ <rho, theta, phi>;
```

$$\begin{bmatrix} \sin(\phi)\cos(\theta) & -\rho\sin(\phi)\sin(\theta) & \rho\cos(\phi)\cos(\theta) \\ \sin(\phi)\sin(\theta) & \rho\sin(\phi)\cos(\theta) & \rho\cos(\phi)\sin(\theta) \\ \cos(\phi) & 0 & -\rho\sin(\phi) \end{bmatrix}$$

$$-\sin(\phi)\,\rho^2$$

Notice that the Jacobian determinant $-\rho^2 \sin(\phi)$ is negative (which says that this spherical coordinate system is left handed). So the Jacobian is its negative

```
> J:=-JDet(R) &@ <rho, theta, phi>;
```

$$J := \sin(\phi)\,\rho^2$$

and the volume element is $dV = \rho^2 \sin(\phi)\, d\rho\, d\theta\, d\phi$. This justifies the formula given in section 5.2.3 and used in example 5.8 which was stated and used there without any real geometrical proof.

In computing a multiple integral, the most important thing is to pick a curvilinear coordinate system adapted to the region and/or the integrand. This is done in the next examples:

Example: 5.17. Compute the integral $\displaystyle\iint_R (x^2 - y^2)\, dA$ over the parallelogram R between the lines $y = x$, $y = x + 2$, $y = 4 - 3x$ and $y = 8 - 3x$.

SOLUTION: To see the region, we first plot the four lines:

```
> plot({x, x+2, 4-3*x, 8-3*x}, x=0..2.5);
```

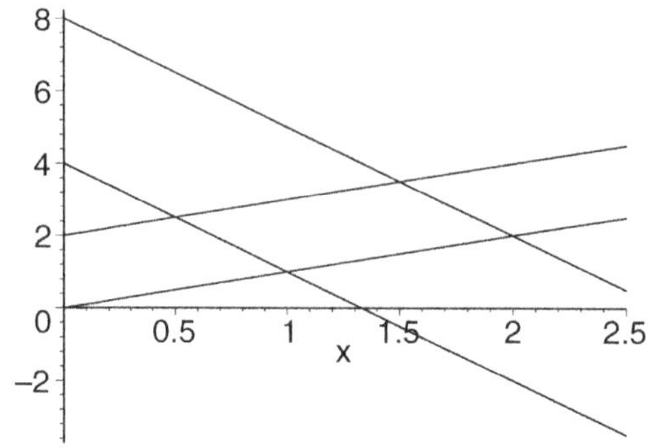

To do the integral, we first want to construct an adapted curvilinear coordinate system. Examining the four lines, we see that if we define $u = y - x$ and $v = y + 3x$ then the four boundaries become $u = 0$, $u = 2$, $v = 4$ and $v = 8$. So we enter these two equations and solve for x and y:

```
> eqs:={ u=y-x, v=y+3*x };
```

$$eqs := \{u = y - x,\ v = y + 3\,x\}$$

```
> sol:=solve(eqs, {x,y});
```

$$sol := \{x = -\frac{u}{4} + \frac{v}{4},\ y = \frac{v}{4} + \frac{3\,u}{4}\}$$

This is the curvilinear coordinate system we will use. It can be converted into a vector of expressions:

```
> Rexp:=subs(sol,<x|y>);
```

$$Rexp := \left[-\frac{u}{4} + \frac{v}{4},\ \frac{v}{4} + \frac{3\,u}{4}\right]$$

and then into a vector of arrow defined functions:

```
> R:=MF(<u|v>, Rexp);
```

$$R := \left[(u,\ v) \to -\frac{1}{4}\,u + \frac{1}{4}\,v,\ (u,\ v) \to \frac{1}{4}\,v + \frac{3}{4}\,u\right]$$

The Jacobian determinant is

```
> JDet(R);
```

$$(u,\ v) \to \frac{-1}{4}$$

Since the Jacobian determinant is negative, the Jacobian is

```
> J:= - JDet(R) &@ <u|v>;
```

$$J := \frac{1}{4}$$

and the area element is $dA = \frac{1}{4}\,du\,dv$. The last thing we need to do before computing the integral is to rewrite the integrand in terms of the curvilinear coordinates

```
> subs(sol, x^2 - y^2); integrand:=simplify(%);
```

$$(-\frac{u}{4}+\frac{v}{4})^2-(\frac{v}{4}+\frac{3u}{4})^2$$

$$integrand := -\frac{1}{2}u^2-\frac{1}{2}v\,u$$

So the integral is (Don't forget the Jacobian.)

```
> Muint(integrand*J, v=4..8, u=0..2); value(%);
```

$$\int_0^2\int_4^8 -\frac{1}{8}u^2-\frac{1}{8}v\,u\,dv\,du$$

$$\frac{-22}{3}$$

Example: 5.18. Compute the volume between the paraboloids $3z = x^2 + y^2$ and $3z = x^2 + y^2 + 4$ above the region between the parabolas $2y = x^2$ and $2y = x^2 + 3$ between $x = -2$ and $x = 2$ by using the curvilinear coordinate system $u = x$, $v = 2y - x^2$ and $w = 3z - x^2 - y^2$.

SOLUTION: To see the solid region, we plot the shadow region in the xy-plane and the upper and lower surfaces over this region:

```
> shadow:= plot3d(0, x=-2..2, y=x^2/2..(x^2+3)/2):
> lower:= plot3d( (x^2+y^2)/3, x=-2..2, y=x^2/2..(x^2+3)/2, color=gray):
> upper:= plot3d( (x^2+y^2+4)/3, x=-2..2, y=x^2/2..(x^2+3)/2):
> display({shadow, lower, upper}, orientation=[75,75],
> axes=normal);
```

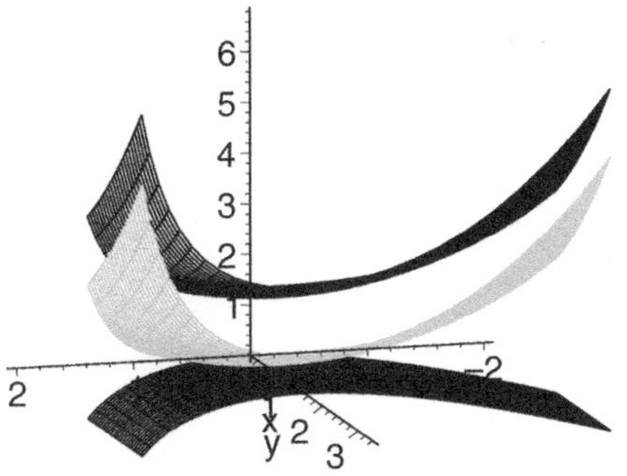

To do the integral, we first enter the equations for the curvilinear coordinates and solve for x, y and z:

```
> eqs:={ u=x, v=2*y-x^2, w=3*z-x^2-y^2 };
```

$$eqs := \{u = x,\ v = 2\,y - x^2,\ w = 3\,z - x^2 - y^2\}$$

```
> sol:=solve(eqs, {x, y, z});
```

$$sol := \{x = u,\ y = \frac{v}{2}+\frac{u^2}{2},\ z = \frac{1}{3}w+\frac{1}{3}u^2+\frac{1}{12}v^2+\frac{1}{6}v\,u^2+\frac{1}{12}u^4\}$$

We then convert this into a vector of arrow-defined functions:

```
> Rexp:=subs(sol,<x|y|z>); R:=MF(<u|v|w>, Rexp);
```

$$Rexp := \left[u, \frac{v}{2} + \frac{u^2}{2}, \frac{1}{3}w + \frac{1}{3}u^2 + \frac{1}{12}v^2 + \frac{1}{6}v\,u^2 + \frac{1}{12}u^4\right]$$

$$R := \left[(u, v, w) \to u, (u, v, w) \to \frac{1}{2}v + \frac{1}{2}u^2, (u, v, w) \to \frac{1}{3}w + \frac{1}{3}u^2 + \frac{1}{12}v^2 + \frac{1}{6}v\,u^2 + \frac{1}{12}u^4\right]$$

The Jacobian determinant is

```
> J:=JDet(R) &@ <u|v|w>;
```

$$J := \frac{1}{6}$$

Since this is positive, it is also the Jacobian factor and the volume element is $dV = \frac{1}{6}\,du\,dv\,dw$. The last thing we need to do before computing the integral is to notice that the boundary equations say the limits are $-2 \le u \le 2$, $0 \le v \le 3$ and $0 \le w \le 4$.

So the volume integral is (Don't forget the Jacobian.)

```
> Muint(1*J, u=-2..2, v=0..3, w=0..4); V:=value(%);
```

$$\int_0^4 \int_0^3 \int_{-2}^2 \frac{1}{6}\,du\,dv\,dw$$

$$V := 8$$

Example: 5.19. Compute the volume inside the cone $z = \sqrt{x^2 + y^2}$ for $0 \le z \le 1$.

SOLUTION: In cylindrical coordinates the cone is given by $z = r$. The piece up to $z = 1$ may be plotted as

```
> plot3d(z, theta=0..2*Pi, z=0..1, axes=normal, coords=cylindrical);
```

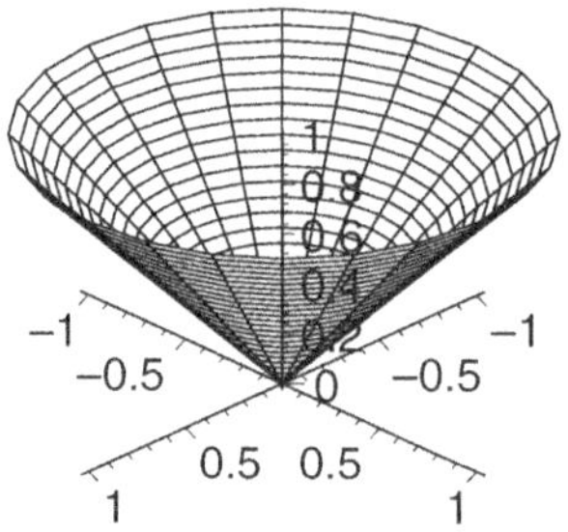

Remember, the default for a cylindrical plot is to give r as a function of θ and z. So the volume inside the cone is

```
> Muint(1*r, r=0..z, theta=0..2*Pi, z=0..1); V:=value(%);
```

$$\int_0^1 \int_0^{2\pi} \int_0^z r\,dr\,d\theta\,dz$$

$$V := \frac{\pi}{3}$$

Of course, this is $V = \frac{1}{3} \times Base \times height = \frac{1}{3} \times \pi(1)^2 \times 1$.

Integrals are not limited to 3-dimensions. The next example generalized the previous example.

Example: 5.20. In $\mathbb{R}^4$, find the 4-dimensional volume inside the 4-dimensional cone which has the rectangular equation $w = \sqrt{x^2 + y^2 + z^2}$ for $0 \le w \le 1$.

SOLUTION: The 4-dimensional generalization of cylindrical coordinates is given by

```
> R:=MF(<rho,theta,phi,w>, <rho*sin(phi)*cos(theta),
> rho*sin(phi)*sin(theta), rho*cos(phi), w>);
```

$$R := \begin{bmatrix} (\rho, \theta, \phi, w) \to \rho \sin(\phi)\cos(\theta) \\ (\rho, \theta, \phi, w) \to \rho \sin(\phi)\sin(\theta) \\ (\rho, \theta, \phi, w) \to \rho \cos(\phi) \\ (\rho, \theta, \phi, w) \to w \end{bmatrix}$$

which is spherical coordinates for x, y and z with an extra w coordinate. Then the equation of the cone is $w = \rho$. We can't plot in 4D but we can still compute the 4-volume. The Jacobian determinant and the Jacobian factor are

```
> JDet(R) &@ <rho,theta,phi,w>; J:= - JDet(R) &@ <rho,theta,phi,w>;
```

$$-\sin(\phi)\,\rho^2$$

$$J := \sin(\phi)\,\rho^2$$

So the volume is

```
> Muint(1*J, rho=0..w, theta=0..2*Pi, phi=0..Pi, w=0..1);
> V:=value(%);
```

$$\int_0^1 \int_0^{\pi} \int_0^{2\pi} \int_0^{w} \sin(\phi)\,\rho^2\, d\rho\, d\theta\, d\phi\, dw$$

$$V := \frac{\pi}{3}$$

You should notice that this is $V = \frac{1}{4} \times Base \times height = \frac{1}{4} \times \frac{4}{3}\pi(1)^3 \times 1$.

Sometimes you need to make two changes of variables, as in the next example.

Example: 5.21. Find the volume below the function $z = e^{-x^2/16 - y^2/9}$ above the region E in the xy-plane enclosed in the ellipse $\frac{x^2}{16} + \frac{y^2}{9} = 25$

SOLUTION: The integrand is

```
> z1:= exp(-x^2/16-y^2/9);
```

$$z1 := e^{(-\frac{x^2}{16} - \frac{y^2}{9})}$$

So the rectangular integral is

```
> Muint(z1,x,y);
```

$$\int\int e^{(-\frac{x^2}{16} - \frac{y^2}{9})}\, dx\, dy$$

over the ellipse E.

We first notice that the formula for the ellipse will be simpler if we define curvilinear coordinates $u = \dfrac{x}{4}$ and $v = \dfrac{y}{3}$ so that the equation of the ellipse becomes $u^2 + v^2 = 25$ which is a circle C of radius 5 in the uv-plane. So we define the curvilinear coordinates

```
> R:=MF(<u|v>,<4*u|3*v>);
```

$$R := [(u,\, v) \to 4\,u,\, (u,\, v) \to 3\,v]$$

compute the Jacobian

```
> JR:=JDet(R) &@ <u|v>;
```

$$JR := 12$$

and redefine the integrand

```
> eqs:=equate(<x|y>,R &@ <u|v>);
```

$$eqs := \{x = 4\,u,\, y = 3\,v\}$$

```
> z2:=subs(eqs, z1);
```

$$z2 := e^{(-u^2-v^2)}$$

So the integral is now

```
> Muint(z2*JR, u, v);
```

$$\int\int 12\,e^{(-u^2-v^2)}\,du\,dv$$

over the circle C.

We now notice that it would be better to do this in polar coordinates (ρ, θ) in the uv-plane. So we define the coordinate transformation

```
> T:=MF(<rho|theta>, <rho*cos(theta)|rho*sin(theta)>);
```

$$T := [(\rho,\, \theta) \to \rho\cos(\theta),\, (\rho,\, \theta) \to \rho\sin(\theta)]$$

compute the Jacobian

```
> JT:=JDet(T) &@ <rho|theta>;
```

$$JT := \rho$$

and redefine the integrand (Don't forget that `JR` is now part of the integrand.)

```
> eqs2:=equate(<u|v>,T &@ <rho|theta>);
```

$$eqs2 := \{u = \rho\cos(\theta),\, v = \rho\sin(\theta)\}$$

```
> z3:=simplify(subs(eqs2, z2*JR));
```

$$z3 := 12\,e^{(-\rho^2)}$$

So the integral is now

```
> Muint(z3*JT, rho, theta);
```

$$\int\int 12\,e^{(-\rho^2)}\,\rho\,d\rho\,d\theta$$

over the region in the $\rho\theta$-plane which is just the rectangle $0 \le \rho \le 5$ and $0 \le \theta \le 2\pi$. So the integral and the final volume are

```
> Muint(z3*JT, rho=0..5, theta=0..2*Pi); V:=value(%);
```

$$\int_0^{2\pi} \int_0^5 12\, e^{(-\rho^2)}\, \rho\, d\rho\, d\theta$$

$$V := 12\,\pi - 12\, e^{(-25)}\, \pi$$

Of course, we could have done this in a single step using the elliptic coordinate system (ρ, θ) defined by $x = 4\rho\cos(\theta)$ and $y = 3\rho\sin(\theta)$ whose Jacobian is 12ρ.

5.4 Exercises

- Do Lab: 9.8.
- Do Projects: 10.10, 10.11 and 10.13.

1. Compute the double integral $\displaystyle\int_{-1}^{1} \int_0^1 (x^2y^4 + 3x^4y^2)\, dy\, dx.$

2. Evaluate the triple integral $\displaystyle\int_0^1 \int_0^z \int_0^y xyz\, dx\, dy\, dz.$

3. Consider the integral $\displaystyle\int_0^4 \int_{x/2}^{\sqrt{x}} x^2y\, dy\, dx.$

   (a) Compute the integral.

   (b) Plot the region of integration in the xy-plane.

   (c) Reverse the order of integration and recompute the integral.

4. Consider the integral $\displaystyle\int_0^{\sqrt{\pi}} \int_y^{\sqrt{\pi}} \sin(x^2)\, dx\, dy$

   (a) On paper, by hand, draw the region of integration in the xy-plane.

   (b) Reverse the order of integration and compute the integral showing all the intermediate steps.

   (c) Return to the original order of integration and compute the integral again showing all the intermediate steps.
   NOTE: FresnelS is a special function that Maple knows about.

   (d) What is the derivative of FresnelS? What is FresnelS(0)?

5. Evaluate the double integral $\displaystyle\int_0^3 \int_{y^2}^9 y\cos(x^2)\, dx\, dy$ explicitly and by reversing the order of integration. Examine the intermediate steps. Which one could you do by hand?

6. Change the order of integration in the integral $\displaystyle\int_0^9 \int_{\sqrt{y}}^3 \sin(\pi x^3)\, dx\, dy.$ Try to compute both integrals. Examine the intermediate steps. Explain what happened.

7. Change the order of integration in the integral $\int_0^1 \int_{\sqrt{y}}^1 \sqrt{x^3+1}\, dx\, dy$. Try to compute both integrals. Examine the intermediate steps. Explain what happened.

8. Find the mass and center of mass of the solid bounded by the parabolic cylinder $y = x^2$ and the two planes given by $z = 0$ (the xy-plane) and $y + z = 4$ (a slanted plane). Here the variable density function is given by $\rho = 1 + x + y + z$.

9. Compute the integral $\iint \sin^6(x^2+y^2)\, dx\, dy$ over the ring between the circles $x^2+y^2 = \frac{\pi}{2}$ and $x^2 + y^2 = \pi$.

10. Find the area of the region inside $r = 4\sin\theta$ and outside $r = 2$. Plot the two curves.

11. Find the area and centroid of the region which lies inside the cardioid $r = 5(1+\cos\theta)$ and outside the circle $r = 5$. Plot the two curves.

12. Find the area and centroid of the region which lies inside the curve $r = 3\cos\theta$ and outside the curve $r = 2 - \cos\theta$. Plot the two curves.

13. Use spherical coordinates to evaluate the triple integral $\iiint_B x^2 + y^2 + z^2\, dV$ where B is the ball $x^2 + y^2 + z^2 \leq 9$.

14. Use cylindrical coordinates to evaluate the triple integral $\iiint_C z e^{x^2+y^2}\, dV$ where C is the cylinder $x^2 + y^2 \leq 4 \quad 0 \leq z \leq 4$.

15. Use cylindrical coordinates to evaluate the triple integral $\iiint_C z e^{x^2+y^2}\, dV$ where C is the cone $\sqrt{x^2+y^2} \leq z \leq 4$.

16. Compute the volume and centroid of the solid bounded on the sides by the circular cylinder $x^2 + y^2 = 4$, below by the plane $z = 0$, and above by the slanted plane $y + z = 3$.

17. Find the mass and center of mass of a solid hemisphere H of radius a whose density at any point is proportional to the distance from the center of the base of the hemisphere.

18. Change the triple integral $\int_{-1}^1 \int_{-\sqrt{1-x^2}}^{\sqrt{1-x^2}} \int_{x^2+y^2}^{2-x^2-y^2} (x^2+y^2)^{3/2}\, dz\, dy\, dx$ into cylindrical coordinates. Try to compute both the cylindrical and rectangular integrals. Examine the intermediate steps. Compare the results using `evalf`. Which one would you prefer to do by hand?

19. Change the triple integral $\int_{-5}^5 \int_0^{\sqrt{25-x^2}} \int_0^{\sqrt{25-x^2-y^2}} \frac{1}{\sqrt{x^2+y^2+z^2}}\, dz\, dy\, dx$ into spherical coordinates. Try to compute both the spherical and rectangular integrals. Examine the intermediate steps. Which one would you prefer to do by hand?

20. Use the transformation $T: \quad x = \frac{2}{3}u + \frac{1}{3}v, \quad y = -\frac{1}{3}u + \frac{1}{3}v$ to evaluate $\displaystyle\iint_Q \frac{x+2y}{\cos(x-y)}\, dA$, where Q is the parallelogram in the xy-plane bounded by the lines $x+2y=0$, $x-y=1$, $x+2y=2$, $x-y=0$. NOTE: The *inverse transformation* is $T^{-1}: \quad u = x-y, \quad v = x+2y$. This is useful in determining the new limits of integration.

21. Use the transformation $T: \quad x = \frac{1}{2}u + \frac{1}{2}v, \quad y = -\frac{1}{2}u + \frac{1}{2}v$ to evaluate $\displaystyle\iint_Q \frac{x-y}{x+y}\, dA$, where Q is the parallelogram in the xy-plane bounded by the lines $x+y=2$, $x-y=0$, $x+y=4$, $x-y=-2$. NOTE: The *inverse transformation* is $T^{-1}: \quad u = x-y, \quad v = x+y$.

22. Plot the four curves $y = \frac{1}{x}$, $y = \frac{2}{x}$, $y = \frac{2}{x^2}$, $y = \frac{4}{x^2}$ for $.5 \le x \le 5$ and $0 \le y \le 5$. Then compute the integral $\displaystyle\iint_R x^2 y\, dx\, dy$ over the "diamond" shaped region bounded by these four curves.

HINT: Define the curvilinear coordinates $u = xy$ and $v = x^2 y$. In terms of u and v, what are the boundary curves? What are the ranges for u and v for the region? Solve for x and y in terms of u and v. Find the Jacobian factor and the integrand. Then integrate.

23. Plot the four curves $y = 1 + \frac{1}{2}e^x$, $y = 2 + \frac{1}{2}e^x$, $y = 3 - \frac{1}{2}e^x$, $y = 6 - \frac{1}{2}e^x$ for $-1 \le x \le 3$ and $0 \le y \le 5$. Then compute the integral $\displaystyle\iint_R y^2 e^{2x}\, dx\, dy$ over the "diamond" shaped region bounded by these four curves.

HINT: Define the curvilinear coordinates $u = y - \frac{1}{2}e^x$ and $v = y + \frac{1}{2}e^x$. In terms of u and v, what are the boundary curves? What are the ranges for u and v for the region? Solve for x and y in terms of u and v. Find the Jacobian factor and the integrand. Then integrate.

Chapter 6

Line and Surface Integrals

6.1 Parametrized Curves

Parametric curves were introducedin section 1.3 and their differential propertieswere discussed in section 2.2. In this section, we will discuss their integral properties

So, consider a parametrized curve whose position vector is given by $\vec{r}(t) = \langle x(t), y(t), z(t) \rangle$. Then its velocity is $\vec{v}(t) = \dfrac{d\vec{r}}{dt} = \left\langle \dfrac{dx}{dt}, \dfrac{dy}{dt}, \dfrac{dz}{dt} \right\rangle$ and its speed is $|\vec{v}(t)| = \sqrt{\dfrac{dx}{dt}^2 + \dfrac{dy}{dt}^2 + \dfrac{dz}{dt}^2}$.

6.1.1 Line Integrals of Scalars

In section 2.2, we used the *(scalar) differential of arc length*,

$$ds = \sqrt{dx^2 + dy^2 + dz^2} = |\vec{v}|\,dt\ ,$$

to compute the arc length from $A = \vec{r}(a)$ to $B = \vec{r}(b)$ as

$$s(A, B) = \int_A^B ds = \int_a^b |\vec{v}|\,dt.$$

It can also be used to define the line integral of a *scalar* function $f(t)$ defined along the curve to be

$$\int_A^B f ds = \int_a^b f(t)|\vec{v}(t)|\,dt\ .$$

Alternatively, if $f(x, y, z)$ is defined throughout space, then it may be restricted to the curve by composing with $\vec{r}(t)$ and then its integral is

$$\int_A^B f ds = \int_a^b f(\vec{r}(t))|\vec{v}(t)|\,dt.$$

Example: 6.1. Plot the spiral helix $\vec{r}(t) = \langle t\cos(t), t\sin(t), t\rangle$ for $0 \le t \le 8\pi$ and compute the integral of the function $f(x, y, z) = z(2 + x^2 + y^2)$ over this portion of the spiral helix.

SOLUTION: We enter the curve and plot it:

```
> r:=MF(t,<t*cos(t)| t*sin(t)| t>);
```

$$r := [t \to t\cos(t),\ t \to t\sin(t),\ t \to t]$$

```
> spacecurve([r[1](t),r[2](t),r[3](t)], t=0..8*Pi, numpoints=96,
> axes=normal, orientation=[30,50], thickness=2);
```

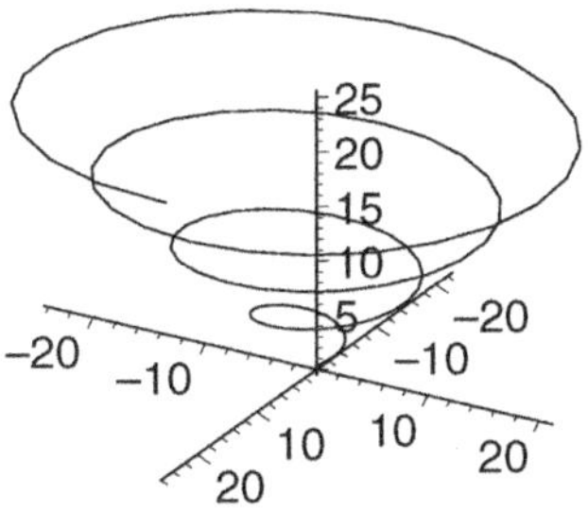

The function $f(x, y, z) = z(2 + x^2 + y^2)$ may be restricted to the spiral helix by forming the composition $f(\vec{r}(t)) = t\left(2 + (t\cos(t))^2 + (t\sin(t))^2\right)$. In *Maple* we enter the function

```
> f:=MF(<x,y,z>, z*(2 + x^2 + y^2));
```

$$f := (x, y, z) \to z\,(2 + x^2 + y^2)$$

form the composition and simplify it.

```
> fr:=simplify(f &@ (r &@ t));
```

$$fr := t\,(2 + t^2)$$

The integral of f between $t = 0$ and $t = 8\pi$ is $\displaystyle\int_{(0,0,0)}^{(8\pi,0,8\pi)} f\,ds = \int_0^{8\pi} f(\vec{r}(t))|\vec{v}|\,dt$. Using *Maple*, the velocity and speed are

```
> v:=D(r);
```

$$v := [t \to \cos(t) - t\sin(t),\ t \to \sin(t) + t\cos(t),\ 1]$$

```
> speed:=Length(v &@ t);
```

$$speed := \sqrt{2 + t^2}$$

So, the integral is

```
> Int(fr * speed, t=0..8*Pi); expand(value(%));
```

$$\int_0^{8\pi} t\,(2 + t^2)^{(3/2)}\,dt$$

$$\frac{32768\,\pi^5\,(1 + \frac{1}{32\,\pi^2})^{(5/2)}}{5} - \frac{4\sqrt{2}}{5}$$

6.1.2 Mass, Center of Mass and Moment of Inertia

Table C.2 in Appendix C, shows the standard applications of line integrals of scalar functions. As examples, we will discuss the mass, center of mass and moment of inertia of a wire with a specified linear density.

Suppose the wire has the shape of a curve $\vec{r}(t)$ and has linear density $\rho(t)$ at the point $\vec{r}(t)$. (Notice that $\rho(t)$ is measured in units of mass per unit length so that $\rho(t)\,ds$ is the mass of a piece of wire of length ds.) Then the mass of the wire between $A = \vec{r}(a)$ and $B = \vec{r}(b)$ is

$$M = \int_A^B \rho\,ds = \int_a^b \rho(t)|\vec{v}|\,dt\ .$$

and the center of mass is

$$(\overline{x}, \overline{y}, \overline{z}) = \left(\frac{M_{yz}}{M}, \frac{M_{xz}}{M}, \frac{M_{xy}}{M}\right)$$

where the first moments are

$$M_{yz} = \int_A^B x\,\rho\,ds = \int_a^b x(t)\,\rho(t)|\vec{v}|\,dt$$
$$M_{xz} = \int_A^B y\,\rho\,ds = \int_a^b y(t)\,\rho(t)|\vec{v}|\,dt$$
$$M_{xy} = \int_A^B z\,\rho\,ds = \int_a^b z(t)\,\rho(t)|\vec{v}|\,dt\ .$$

Example: 6.2. Suppose a wire has the shape of the spiral helix of example 6.1, and has density proportional to the distance from the xy-plane. Find its mass and center of mass.

SOLUTION: The distance from the xy-plane is z. So the density is $\rho = Kz$ (for some constant K), which may be entered as

```
> rho:=K*r[3](t);
```

$$\rho := K\,t$$

Hence the mass is

```
> Int(rho * speed, t=0..8*Pi); M:=value(%);
```

$$\int_0^{8\,\pi} K\,t\,\sqrt{2+t^2}\,dt$$

$$M := -\frac{2\,\sqrt{2}\,K}{3} + \frac{(2+64\,\pi^2)^{(3/2)}\,K}{3}$$

To find the center of mass, we first compute the moments. *Maple* is unable to compute the integrals exactly using `value`, for example:

```
> Int(r[1](t) * rho * speed, t=0..8*Pi); Myz:=value(%);
```

$$\int_0^{8\,\pi} t^2\cos(t)\,K\,\sqrt{2+t^2}\,dt$$

$$Myz := \int_0^{8\,\pi} t^2\cos(t)\,K\,\sqrt{2+t^2}\,dt$$

So we get an approximate value using `evalf`:

NOTE: The command `expand` is needed to factor out the constant K.

```
> Myz:=evalf(expand(Myz));
```

$$Myz := 1894.954419\, K$$

The other two moments are

```
> Int(r[2](t) * rho * speed, t=0..8*Pi); Mxz:=evalf(expand(%));
```

$$\int_0^{8\,\pi} t^2 \sin(t)\, K\, \sqrt{2+t^2}\, dt$$

$$Mxz := -15749.84329\, K$$

```
> Int(r[3](t) * rho * speed, t=0..8*Pi); Mxy:=evalf(expand(%));
```

$$\int_0^{8\,\pi} t^2\, K\, \sqrt{2+t^2}\, dt$$

$$Mxy := 100061.0758\, K$$

and the center of mass is

```
> CM:=evalf(<Myz/M| Mxz/M| Mxy/M>);
```

$$CM := [0.3564659633,\ -2.962753618,\ 18.82281041]$$

Examine the plot in example 6.1 and notice that the center of mass is near the z-axis but above the center since the density of the spiral helix is greater toward the top.

The moments of inertia about the three axes are:

$$I_x = \int_A^B (y^2+z^2)\,\rho\, ds = \int_a^b (y(t)^2+z(t)^2)\,\rho(t)|\vec{v}|\, dt$$
$$I_y = \int_A^B (x^2+z^2)\,\rho\, ds = \int_a^b (x(t)^2+z(t)^2)\,\rho(t)|\vec{v}|\, dt$$
$$I_z = \int_A^B (x^2+y^2)\,\rho\, ds = \int_a^b (x(t)^2+y(t)^2)\,\rho(t)|\vec{v}|\, dt$$

Example: 6.3. Find the moments of inertia of a wire in the shape of the spiral helix of example 6.1 with density proportional to the distance from the xy-plane.

SOLUTION: The quantities were all defined in the previous examples. So the moments of inertia are

```
> Int( (r[2](t)^2 + r[3](t)^2) * rho * speed, t=0..8*Pi);
> Ix:=evalf(expand(%));
```

$$\int_0^{8\,\pi} (t^2 \sin(t)^2 + t^2)\, K\, t\, \sqrt{2+t^2}\, dt$$

$$Ix := 0.3008290674\, 10^7\, K$$

```
> Int( (r[1](t)^2 + r[3](t)^2) * rho * speed, t=0..8*Pi);
> Iy:=evalf(expand(%));
```

$$\int_0^{8\pi} (t^2 \cos(t)^2 + t^2)\, K\, t\, \sqrt{2+t^2}\, dt$$

$$Iy := 0.3024140862\, 10^7\, K$$

```
> Int( (r[1](t)^2 + r[2](t)^2) * rho * speed, t=0..8*Pi);
> Iz:=evalf(expand(%));
```

$$\int_0^{8\pi} (t^2 \cos(t)^2 + t^2 \sin(t)^2)\, K\, t\, \sqrt{2+t^2}\, dt$$

$$Iz := 0.2010810512\, 10^7\, K$$

6.1.3 Line Integrals of Vectors

Given a vector field

$$\vec{F}(x,y,z) = \langle F_1(x,y,z), F_2(x,y,z), F_3(x,y,z)\rangle,$$

the line integral of $\vec{F}$ over a curve $\vec{r}(t)$ is

$$\int_A^B \vec{F}\cdot d\vec{s} = \int_a^b \vec{F}(\vec{r}(t))\cdot\vec{v}(t)\, dt = \int_A^B \vec{F}\cdot\hat{T}\, ds$$

where the *vector differential of arc length* is:

$$d\vec{s} = \langle dx, dy, dz\rangle = \left\langle \frac{dx}{dt}, \frac{dy}{dt}, \frac{dz}{dt}\right\rangle dt = \vec{v}\, dt = \hat{T}\, ds\ .$$

and where $\vec{F}(\vec{r}(t))$ is the composition of the vector field $\vec{F}(x,y,z)$ and the curve $\vec{r}(t)$. We will also say that $\vec{F}(\vec{r}(t))$ is the restriction of $\vec{F}$ to the curve or the value of $\vec{F}$ along the curve. Writing the integral in the form $\int_A^B \vec{F}\cdot\hat{T}\, ds$ with the unit tangent vector $\hat{T}$, is useful for theoretical purposes, but it is more convenient to compute it in the form $\int_a^b \vec{F}(\vec{r}(t))\cdot\vec{v}(t)\, dt$.

Example: 6.4. Compute the line integral $\int_{(0,0,0)}^{(6\pi,0,6\pi)} \vec{F}\cdot d\vec{s}$ of the vector field $\vec{F} = (3xz, 2xy, x^2)$ along the spiral helix $\vec{r}(t) = (t\cos(t), t\sin(t), t)$. Then plot the vector field and the spiral helix in the same plot.

SOLUTION: We input the curve and the vector field:

```
> r:=MF(t, <t*cos(t)| t*sin(t)| t>);
```

$$r := [t \to t\cos(t),\ t \to t\sin(t),\ t \to t]$$

```
> F:=MF(<x|y|z>, <3*x*z| 2*x*y| x^2>);
```

$$F := [(x, y, z) \to 3\, z\, x,\ (x, y, z) \to 2\, y\, x,\ (x, y, z) \to x^2]$$

Then we compute the velocity and evaluate the vector field on the curve:

```
> v:=D(r);
```

$$v := [t \to \cos(t) - t\sin(t),\ t \to \sin(t) + t\cos(t),\ 1]$$

```
> Fr:=F &@ r &@ t;
```

$$Fr := [3\,t^2\cos(t),\ 2\,t^2\sin(t)\cos(t),\ t^2\cos(t)^2]$$

Next we compare the endpoints $(0,0,0)$ and $(6\pi,0,6\pi)$ with (the z-component of) the parametrization $(t\cos(t), t\sin(t), t)$ and observe that the parameter range is $0 \le t \le 6\pi$. Hence the integral is

```
> Int(Fr &.  (v &@ t), t=0..6*Pi); value(%);
```

$$\int_0^{6\pi} -t^2\cos(t)\,(-2\sin(t)\,t\cos(t) - 4\cos(t) + 3\,t\sin(t) - 2 + 2\cos(t)^2)\,dt$$

$$\frac{93}{4}\pi + 162\,\pi^3$$

Finally, we plot the spiral helix and the vector field and display them in the same plot:

```
> pr:=spacecurve([r[1](t),r[2](t),r[3](t)], t=0..6*Pi, numpoints=96,
> thickness=2):
> pF:=fieldplot3d(F &@ <x,y,z>, x=-6*Pi..6*Pi, y=-6*Pi..6*Pi,
> z=0..6*Pi):
> display({pr,pF}, axes=normal, orientation=[30,50]);
```

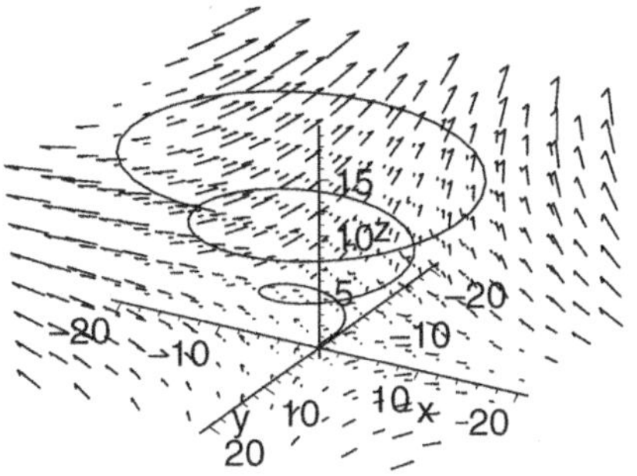

The line integral of a vector field can also be written as:

$$\int_A^B \vec{F}\cdot d\vec{s} = \int_A^B F_1(x,y,z)\,dx + F_2(x,y,z)\,dy + F_3(x,y,z)\,dz$$

In this form, the integral is computed (by hand) by replacing the coordinates and the differentials by their values on the curve and then integrating with respect to the parameter. However, on the computer, it is still easier to integrate the dot product of the vector field and the velocity.

Example: 6.5. Compute the line integral $\int_{(0,0,0)}^{(-3\pi,0,3\pi)} -x^2y\,dx + y^2x\,dy + z^3\,dz$ along the spiral helix $\vec{r}(t) = (t\cos(t), t\sin(t), t)$.

SOLUTION: We notice that the integral is $\int_{(0,0,0)}^{(-3\pi,0,3\pi)} \vec{F} \cdot d\vec{s}$ for the vector field $\vec{F} = (-x^2y, y^2x, z^3)$

So we enter the vector field:

```
> F:=MF(<x,y,z>, <-x^2*y| y^2*x| z^3>);
```

$$F := [(x,\, y,\, z) \to -x^2\, y,\; (x,\, y,\, z) \to y^2\, x,\; (x,\, y,\, z) \to z^3]$$

and evaluate the vector field on the curve:

```
> Fr:=F &@ r &@ t;
```

$$Fr := \left[-t^3 \cos(t)^2 \sin(t),\; t^3 \sin(t)^2 \cos(t),\; t^3\right]$$

Next we compare the endpoints with (the z-component of) the parametrization and conclude that the parameter range is $0 \le t \le 3\pi$. Hence the integral is

```
> Int(Fr &.  (v &@ t), t=0..3*Pi); value(%);
```

$$\int_0^{3\,\pi} -t^3\,(-\sin(t)\cos(t) + 2\cos(t)^3 \sin(t) + 2\cos(t)^4\, t - 1 - 2\cos(t)^2\, t)\, dt$$

$$\frac{81}{4}\,\pi^4 + \frac{243}{20}\,\pi^5$$

6.1.4 Work and Circulation

Table C.3 in Appendix C, shows the standard applications of line integrals of vector functions. As examples, we consider the work a force does on a particle and the circulation of a fluid (or of an electric or magnetic field).

Work If a particle moves along a curve $\vec{r}(t)$ due to the action of a force $\vec{F}(x, y, z)$, then the work done on the particle is the line integral of the tangential component of the force along the curve:

$$\mathcal{W}ork = \int_A^B \vec{F} \cdot \hat{T}\, ds = \int_A^B \vec{F} \cdot d\vec{s}\,.$$

Example: 6.6. A 57 kg satellite is falling out of orbit along the spiral curve $R(t) = \left((7124 - 13t)\cos\left(\frac{2\pi t}{87}\right), (7124 - 13t)\sin\left(\frac{2\pi t}{87}\right), 0\right)$ where t is in minutes. Find the work done on the satellite by the gravitational force $\vec{F} = -\frac{GMm}{r^3}\vec{r}$ as the satellite falls from an altitude of 7124 km (measured from the center of the earth) to the earth's surface at 6371 km.

SOLUTION: We first input the curve and compute the velocity:

```
> R:=MF(t,<(7124-13*t) * cos(2*Pi*t/87)| (7124-13*t) * sin(2*Pi*t/87)|
> 0>);
```

$$R := \left[t \to -13\,(-548 + t)\cos\left(\frac{2}{87}\, t\, \pi\right),\; t \to -13\,(-548 + t)\sin\left(\frac{2}{87}\, t\, \pi\right),\; t \to 0\right]$$

```
> V:=D(R);
```

$$V := \left[t \to -13\cos(\frac{2}{87}\,t\,\pi) + \frac{26}{87}\,(-548+t)\sin(\frac{2}{87}\,t\,\pi)\,\pi,\right.$$
$$\left. t \to -13\sin(\frac{2}{87}\,t\,\pi) - \frac{26}{87}\,(-548+t)\cos(\frac{2}{87}\,t\,\pi)\,\pi,\ 0\right]$$

To find the range for the parameter, we compare the curve to the formula for cylindrical coordinates $(r\cos\theta, r\sin\theta, z)$. So the altitude (radius from the center of the earth) is $r = 7124 - 13t$. Thus, the altitude is 7124 km at $t = 0$. We solve for the time when the altitude is 6371 km:

```
> t2:=fsolve( 7124 - 13*t = 6371, t);
```

$$t2 := 57.92307692$$

In the force equation, the gravitational constant is $G = 6.67 \times 10^{-11}$ m^3/kg/sec^2, the mass of the earth is $M = 5.97 \times 10^{24}$ kg, the mass of the satellite is $m = 57$ kg, the vector from the center of the earth is $\vec{r} = (x, y, z)$ and the distance from the center of the earth is $r = \sqrt{x^2+y^2+z^2}$. We enter the constants and compute the force:

```
> G:=6.67 * 10^(-11):  M:=5.97 * 10^24:  m:=57:
> F:=MF(<x,y,z>, simplify( G*M*m * <x|y|z> / sqrt(x^2+y^2+z^2)^3 ));
```

$$F := \left[(x,\,y,\,z) \to \frac{0.2269734300\,10^{17}\,x}{(x^2+y^2+z^2)^{(3/2)}},\ (x,\,y,\,z) \to \frac{0.2269734300\,10^{17}\,y}{(x^2+y^2+z^2)^{(3/2)}},\right.$$
$$\left.(x,\,y,\,z) \to \frac{0.2269734300\,10^{17}\,z}{(x^2+y^2+z^2)^{(3/2)}}\right]$$

We then evaluate the force on the curve:

```
> assume(7124-13*t>0);
> FR:=simplify(F &@ R &@ t);
```

$$FR := \left[\frac{0.1343038047\,10^{15}\cos(0.07222052079\,t\tilde{})}{(-548. + t\tilde{})^2},\right.$$
$$\left.\frac{0.1343038047\,10^{15}\sin(0.07222052079\,t\tilde{})}{(-548. + t\tilde{})^2},\ 0.\right]$$

NOTE: To facilitate the simplification, we have assumed that the altitude is positive.

So the work integral is

```
> Int(FR &.  (V &@ t), t=0..t2);  W:=value(%);
```

$$\int_0^{57.92307692} -\frac{0.1745949461\,10^{16}}{(-548. + t\tilde{})^2}\,dt\tilde{}$$

$$W := -0.3765637246\,10^{12}$$

Circulation The instantaneous motion of a fluid is measured by its velocity field $\vec{V}_f(x, y, z)$ which gives the velocity of the fluid at the point $\langle x, y, z\rangle$. The line integral of the tangential component of the velocity field around a closed curve $\vec{r}(t)$ is called the circulation of the fluid around the curve and measures the net flow of the fluid around the curve:

$$Circulation = \oint_{\vec{r}(t)} \vec{V}_f \cdot \hat{T}\, ds = \oint_{\vec{r}(t)} \vec{V}_f \cdot d\vec{s}$$

NOTE: There are two velocities here: the velocity field of the fluid denoted by $\vec{V}_f$ and the velocity of the curve denoted by $\vec{v}$.

It is also common to compute the circulation of an electric field $\oint \vec{E} \cdot d\vec{s}$ or of a magnetic field $\oint \vec{B} \cdot d\vec{s}$.

Example: 6.7. Plot the fluid velocity field $\vec{V}_f = (-y, x)$ and compute its circulation around the two families of circles $x^2 + y^2 = a^2$ and $(x-2)^2 + y^2 = a^2$.

SOLUTION: We input the velocity vector field and plot it:

```
> Vf:=MF(<x,y>, <-y|x>);
```

$$Vf := [(x,\, y) \to -y,\, (x,\, y) \to x]$$

```
> fieldplot(Vf &@ <x,y>, x=-5..5, y=-5..5, scaling=constrained);
```

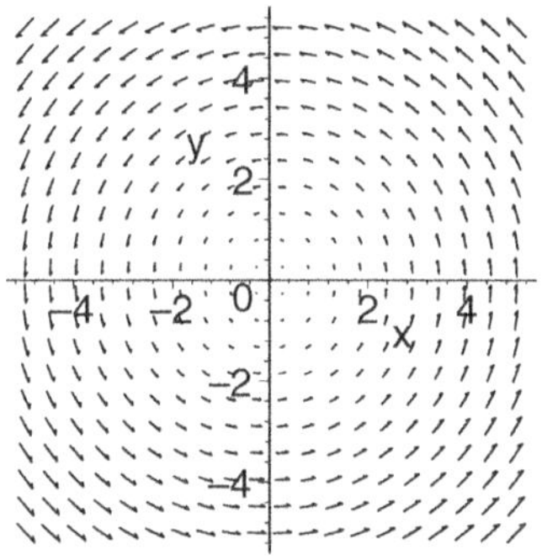

Notice that fluid seems to circulate around the origin.

The first family of circles are centered at the origin. We enter a parametrization and compute the velocity of the curve:

```
> r:=MF(t, <a*cos(t)|a*sin(t)>);
```

$$r := [t \to a\cos(t),\, t \to a\sin(t)]$$

```
> v:=D(r);
```

$$v := [t \to -a\sin(t),\, t \to a\cos(t)]$$

The restriction of the fluid velocity to the curve is

```
> Vfr:=Vf &@ r &@ t;
```

$$Vfr := [-a\sin(t),\, a\cos(t)]$$

and hence the circulation is

```
> Int(Vfr &.  (v &@ t), t=0..2*Pi); C:=value(%);
```

$$\int_0^{2\pi} a^2\, dt$$

$$C := 2\,a^2\,\pi$$

The second family of circles are centered at the point $(2, 0)$. We enter a parametrization and compute the velocity of the curve:

```
> r:=MF(t, <2+a*cos(t)|a*sin(t)>);
```

$$r := [t \to 2 + a\cos(t),\; t \to a\sin(t)]$$

```
> v:=D(r);
```

$$v := [t \to -a\sin(t),\; t \to a\cos(t)]$$

The restriction of the fluid velocity to the curve is

```
> Vfr:=Vf &@ r &@ t;
```

$$Vfr := [-a\sin(t),\; 2 + a\cos(t)]$$

and hence the circulation is

```
> Int(Vfr &. (v &@ t), t=0..2*Pi); C:=value(%);
```

$$\int_0^{2\pi} a\,(2\cos(t) + a)\,dt$$

$$C := 2\,a^2\,\pi$$

Notice that for both families of circles, the circulation is twice the area of the circle. This is not a coincidence and will be explained in subsection 8.3.2 using Green's Theorem.

6.2 Parametrized Surfaces

In $\mathbb{R}^n$, a parametric curve has one parameter, a parametric surface has two parameters and a curvilinear coordinate system has n parameters. In sections 1.3, 2.2 and 6.1, we studied parametrized curves and their differential and integral properties.Similarly, in sections 5.3 and 5.3, we studied curvilinear coordinate systems and their differential and integral properties.

In section 1.3, we introduced parametrized surfaces. We now study their differential and integral properties.These properties are analogous to those for curves and curvilinear coordinate systems. We will restrict attention to surfaces in $\mathbb{R}^3$ but point out those properties which don't generalize to $\mathbb{R}^n$ because they depend on the cross product.

6.2.1 Tangent and Normal Vectors

A surface is specified by giving a list of 3 functions of 2 variables which give the 3 rectangular coordinates as functions of the 2 parameters or coordinates. Thus,

$$(x, y, z) = \vec{R}(u, v) = \big(x(u, v), y(u, v), z(u, v)\big)\ .$$

A coordinate curve is the curve obtained by allowing one parameter to vary while the other parameter is held fixed. If you draw several coordinate curves for each parameter, you obtain a coordinate grid for the surface. The tangent vector to a coordinate curve is obtained by differentiating with respect to the parameter which is varying:

$$\vec{R}_u = \frac{\partial \vec{R}}{\partial u} \qquad \text{and} \qquad \vec{R}_v = \frac{\partial \vec{R}}{\partial v}$$

In $\mathbb{R}^3$, there is one further vector which can be computed. It is the normal vector, which is perpendicular to the surface and may be computed as the cross product of the two coordinate tangent vectors:

$$\vec{N} = \vec{R}_u \times \vec{R}_v$$

A surface may be entered into *Maple* using the `VecCalc` command `MakeFunction` or its alias `MF`. You can plot a parametric surface with its coordinate grid by using the `plot3d` command with a parametric argument. The coordinate tangent vectors may be computed using `D` and the normal may be computed using the `VecCalc` command `Cross` or the `VecCalc` operator `&x`.

Example: 6.8. Plot the spiral ramp $\vec{R}(r,\theta) = \langle r\cos(\theta), r\sin(\theta), \theta\rangle$ for $0 \le r \le 9$ and $0 \le \theta \le 8\pi$. Then compute the coordinate tangent vectors $\vec{R}_r$ and $\vec{R}_\theta$ and the normal vector $\vec{N}$.

SOLUTION: We enter the surface and plot it:

```
> R:=MF(<r,theta>, <r*cos(theta)| r*sin(theta)| theta>);
```

$$R := [(r, \theta) \to r\cos(\theta), (r, \theta) \to r\sin(\theta), (r, \theta) \to \theta]$$

```
> plot3d( R &@ <r,theta>, r=0..9, theta=0..8*Pi, grid=[10,97],
> scaling=constrained, orientation=[45,65]);
```

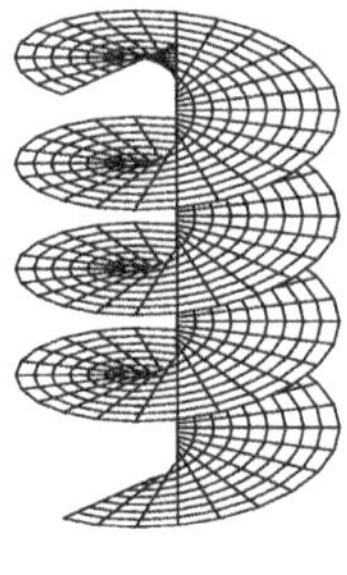

From the plot, we see why the surface is called a spiral ramp.

We next compute the coordinate tangent vectors and the normal vector.

```
> Rr:=D[1](R); Rtheta:=D[2](R);
```

$$Rr := [(r, \theta) \to \cos(\theta), (r, \theta) \to \sin(\theta), 0]$$

$$Rtheta := [(r, \theta) \to -r\sin(\theta), (r, \theta) \to r\cos(\theta), 1]$$

```
> N:=(Rr &@ <r,theta>) &x (Rtheta &@ <r,theta>);
```

$$N := [\sin(\theta), -\cos(\theta), r]$$

NOTE: The cross product operator `&x` expects to act on expressions. So we must evaluate the tangent vectors at the point `<r,theta>` before taking the cross product.

For future reference, notice that the normal points basically upward since the z component $N_z = r$ is positive.

6.2.2 Surface Area

The *(scalar) differential of surface area* is

$$dS = \left|\vec{N}\right| \, du \, dv$$

Here the length of the normal $|\vec{N}|$ plays the role of the Jacobian for the parametric surface. Hence, the area of a region R on a parametric surface is given by

$$A = \iint_R dS = \iint_R \left|\vec{N}\right| \, du \, dv$$

where the limits on the integrals must be taken as the appropriate ranges for the parameters u and v. Notice that these formulas are analogous to those for a curve giving the scalar differential of arc length and the arc length in terms of the length of the velocity.

Example: 6.9. Find the area of one cycle of the spiral ramp of example 6.8 for $0 \le r \le 9$.

SOLUTION: We first compute the length of the normal:

```
> lenN:=Length(N);
```

$$lenN := \sqrt{r^2 + 1}$$

One cycle of the ramp occurs for $0 \le \theta \le 2\pi$. So, the area of the ramp is

```
> Muint(lenN, r=0..9, theta=0..2*Pi); A:=value(%); evalf(%);
```

$$\int_0^{2\pi} \int_0^9 \sqrt{r^2 + 1} \, dr \, d\theta$$

$$A := 9\sqrt{82}\,\pi - \ln(-9 + \sqrt{82})\,\pi$$

$$265.1250053$$

This is slightly larger than the area of a circle of radius 9: $\quad A = \pi 9^2 = 81\pi \approx 254$.

6.2.3 Surface Integrals of Scalars

The *scalar* differential of surface area, dS, can also be used to define the surface integral of a *scalar* function $f(u, v)$ defined along a surface R to be

$$\iint_R f dS = \iint_R f(u, v)\,|\vec{N}|\,du\,dv \; .$$

Alternatively, if $f(x, y, z)$ is defined throughout space, then it may be restricted to the surface by composing with $\vec{R}(u, v)$. Then its integral is

$$\iint_R f dS = \iint_R f(\vec{R}(u,v))\,|\vec{N}|\,du\,dv.$$

Example: 6.10. Compute the integral of the function $f(x, y, z) = x + y - z$ over the spiral ramp of example 6.8 for $0 \le r \le 9$ and $0 \le \theta \le \pi$.

SOLUTION: The function f may be entered as

```
> f:=MF(<x,y,z>, x + y - z);
```

$$f := (x, y, z) \to x + y - z$$

and its restriction to the spiral ramp $f(\vec{R}(r, \theta))$ is:

```
> fR:=f &@ R &@ <r,theta>;
```

$$fR := r\cos(\theta) + r\sin(\theta) - \theta$$

Then the integral of f over the spiral ramp is $\iint_R f\,dS = \int_0^{\pi}\int_0^{9} f(\vec{R}(r,\theta))|\vec{N}|\,dr\,d\theta$, which may be computed using *Maple* as

```
> Muint(fR * lenN, r=0..9, theta=0..Pi); value(%); evalf(%);
```

$$\int_0^{\pi}\int_0^{9} (r\cos(\theta) + r\sin(\theta) - \theta)\sqrt{r^2+1}\,dr\,d\theta$$

$$-\frac{2}{3} + \frac{164\sqrt{82}}{3} - \frac{9\sqrt{82}\,\pi^2}{4} + \frac{1}{4}\ln(-9+\sqrt{82})\,\pi^2$$

$$286.1323620$$

6.2.4 Mass, Center of Mass and Moment of Inertia

Table C.2 in Appendix C, shows the standard applications of surface integrals of scalar functions. As examples, we will discuss the mass, center of mass and moment of inertia of a region of a surface with a specified surface density.

Suppose a sheet of plastic has the shape of a region R on a surface $\vec{R}(u, v)$ and has surface density $\rho(u, v)$ at the point $\vec{R}(u, v)$. (Notice that $\rho(u, v)$ is measured in units of mass per unit area so that $\rho(u, v)\,dS$ is the mass of a piece of the surface with area dS.) Then the mass of the plastic sheet is

$$M = \iint_R \rho\,dS = \iint_R \rho(u,v)\,|\vec{N}|\,du\,dv.$$

Example: 6.11. Suppose the dome of a grain silo has the shape of the paraboloid $z = 30 - \dfrac{x^2+y^2}{20}$ for $z \ge 10$ and has a density $\rho = 320 + \dfrac{x^2+y^2}{10}$ so that the dome is thicker at the bottom. Plot the dome and find the total mass of the dome.

SOLUTION: The surface may be parametrized in polar (or cylindrical) coordinates as

```
> R:=MF(<r,theta>, <r*cos(theta)| r*sin(theta)| 30 - r^2/20>);
```

$$R := \left[(r,\,\theta) \to r\cos(\theta),\,(r,\,\theta) \to r\sin(\theta),\,(r,\,\theta) \to 30 - \frac{1}{20}r^2\right]$$

The bottom of the dome is at $z = 10 = 30 - \dfrac{r^2}{20}$ or at $r = 20$. So we plot the dome and save it for future use:

```
> pdome:=plot3d(R &@ <r,theta>, r=0..20, theta=0..2*Pi, axes=normal,
> scaling=constrained, view=[-20..20, -20..20, 0..30],
> orientation=[30,80]): pdome;
```

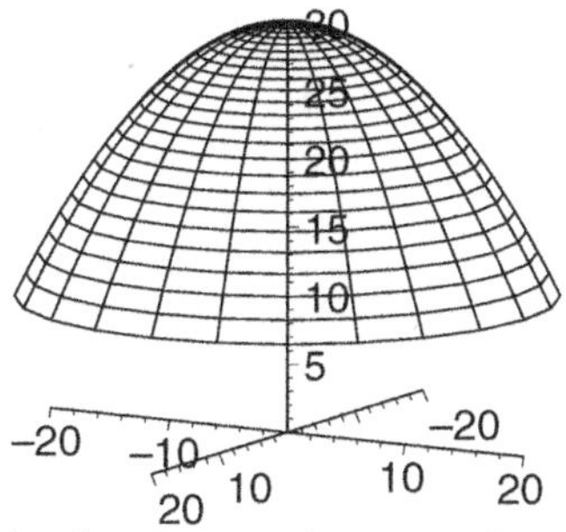

In these coordinates, the density may be entered as

```
> rho := 320 + r^2/10:
```

The coordinate tangent vectors are

```
> Rr:=D[1](R); Rtheta:=D[2](R);
```

$$Rr := \left[(r,\,\theta) \to \cos(\theta),\,(r,\,\theta) \to \sin(\theta),\,(r,\,\theta) \to -\frac{1}{10}r\right]$$

$$Rtheta := [(r,\,\theta) \to -r\sin(\theta),\,(r,\,\theta) \to r\cos(\theta),\,0]$$

The coordinate normal vector and its length are

```
> N:=Cross(Rr &@ <r,theta>, Rtheta &@ <r,theta>); lenN:=Length(N);
```

$$N := \left[\frac{1}{10}r^2\cos(\theta),\,\frac{1}{10}r^2\sin(\theta),\,r\right]$$

$$lenN := \frac{1}{10}\sqrt{100 + r^2}\,|r|$$

So the mass is

```
> Muint(rho * lenN, r=0..20, theta=0..2*Pi); M:=value(%);
```

$$\int_0^{2\pi}\int_0^{20} \frac{1}{10}\,(320 + \frac{r^2}{10})\,\sqrt{100 + r^2}\,|r|\,dr\,d\theta$$

$$M := -\frac{63200}{3}\pi + \frac{340000}{3}\sqrt{5}\,\pi$$

To find the center of mass of a surface, we need to find the first moments:

$$M_{yz} = \iint\limits_R x\,\rho\,dS = \iint\limits_R x(u,v)\rho(u,v)\,|\vec{N}|\,du\,dv$$
$$M_{xz} = \iint\limits_R y\,\rho\,dS = \iint\limits_R y(u,v)\rho(u,v)\,|\vec{N}|\,du\,dv$$
$$M_{xy} = \iint\limits_R z\,\rho\,dS = \iint\limits_R z(u,v)\rho(u,v)\,|\vec{N}|\,du\,dv$$

Then the center of mass is

$$(\overline{x}, \overline{y}, \overline{z}) = \left(\frac{M_{yz}}{M}, \frac{M_{xz}}{M}, \frac{M_{xy}}{M}\right)$$

Example: 6.12. Find the center of mass of the grain silo dome of example 6.11.

SOLUTION: By symmetry, $\overline{x} = 0$ and $\overline{y} = 0$. To find $\overline{z}$, we first find the z-moment

```
> Muint(R[3](r,theta) * rho * lenN, r=0..20, theta=0..2*Pi);
> Mxy:=value(%);
```

$$\int_0^{2\pi}\int_0^{20} \frac{1}{10}\,(30 - \frac{r^2}{20})\,(320 + \frac{r^2}{10})\,\sqrt{100 + r^2}\,|r|\,dr\,d\theta$$

$$Mxy := -\frac{14152000}{21}\,\pi + \frac{47000000}{21}\,\sqrt{5}\,\pi$$

Then the z-component of the center of mass is

```
> zbar:=simplify(Mxy/M); evalf(%);
```

$$zbar := \frac{10\,(-1769 + 5875\,\sqrt{5})}{7\,(-79 + 425\,\sqrt{5})}$$

$$18.63803256$$

Notice that the center of mass is slightly below the center of the height of the dome.

Finally the moments of inertia about the 3 axes are:

$$I_x = \iint\limits_R (y^2 + z^2)\,\rho\,dS = \iint\limits_R (y(u,v)^2 + z(u,v)^2)\rho(u,v)\,|\vec{N}|\,du\,dv$$
$$I_y = \iint\limits_R (x^2 + z^2)\,\rho\,dS = \iint\limits_R (x(u,v)^2 + z(u,v)^2)\rho(u,v)\,|\vec{N}|\,du\,dv$$
$$I_z = \iint\limits_R (x^2 + y^2)\,\rho\,dS = \iint\limits_R (x(u,v)^2 + y(u,v)^2)\rho(u,v)\,|\vec{N}|\,du\,dv$$

Example: 6.13. Find the moment of inertia about the z-axis of the grain silo dome of example 6.11.

SOLUTION: The moment of inertia about the z-axis is

```
> Muint( (R[1](r,theta)^2 + R[2](r,theta)^2) * rho * lenN, r=0..20,
> theta=0..2*Pi); Iz:=value(%);
```

$$\int_0^{2\pi}\int_0^{20}\frac{1}{10}\,(r^2\cos(\theta)^2+r^2\sin(\theta)^2)\,(320+\frac{r^2}{10})\,\sqrt{100+r^2}\,|r|\,dr\,d\theta$$

$$Iz := \frac{17600000}{21}\,\pi+\frac{488000000}{21}\,\sqrt{5}\,\pi$$

6.2.5 Surface Integrals of Vectors

Given a vector field

$$\vec{F}(x,y,z)=\langle F_1(x,y,z),F_2(x,y,z),F_3(x,y,z)\rangle,$$

the surface integral of $\vec{F}$ over a surface $\vec{R}(u,v)$ is

$$\iint_R \vec{F}\cdot d\vec{S}=\iint_R \vec{F}(\vec{R}(u,v))\cdot\vec{N}\,du\,dv=\iint_R \vec{F}\cdot\hat{N}\,dS$$

where the *vector differential of surface area* is:

$$d\vec{S}=\langle dy\,dz,dz\,dx,dx\,dy\rangle=\left\langle\left|\frac{\partial(y,z)}{\partial(u,v)}\right|,\left|\frac{\partial(z,x)}{\partial(u,v)}\right|,\left|\frac{\partial(x,y)}{\partial(u,v)}\right|\right\rangle du\,dv=\vec{N}\,du\,dv=\hat{N}\,dS\ .$$

and where $\vec{F}(\vec{R}(u,v))$ is the composition of the vector field $\vec{F}(x,y,z)$ and the surface $\vec{R}(u,v)$. We will also say that $\vec{F}(\vec{R}(u,v))$ is the restriction of $\vec{F}$ to the surface or the value of $\vec{F}$ along the surface. Writing the integral in the form $\iint_R \vec{F}\cdot\hat{N}\,dS$ with the unit normal vector $\hat{N}$, is useful for theoretical purposes, but it is more convenient to compute it in the form $\iint_R \vec{F}(\vec{R}(u,v))\cdot\vec{N}\,du\,dv$.

It should be noticed that if you reverse the direction of the normal vector $\vec{N}$ then the integral of any vector field changes sign. This means that if you plan to integrate a vector field over a surface then you must specify the side of the surface to which the normal should point and only use a parametrization for which the normal points to the correct side of the surface. However, if you pick a parametrization and find that the normal is backwards, you may correct the problem by reversing the normal by multiplying by -1.

Example: 6.14. Plot the vector field $\vec{V}(x,y,z)=(yz,xz,-z^2)$ and the grain silo dome of example 6.11. (Think of the vector field as rain falling on the dome.) Then compute the integral of $\vec{V}$ over the surface of the grain silo dome with normal pointing downward.

SOLUTION: The vector field may be entered as

```
> V:=MF(<x,y,z>, <y*z| x*z| -z^2>);
```

$$V := [(x, y, z) \to y\,z, (x, y, z) \to z\,x, (x, y, z) \to -z^2]$$

Then we plot it and `display` it with the dome (previously plotted in example 6.11).

```
> pV:= fieldplot3d(V &@ <x,y,z>, x=-20..20, y=-20..20, z=0..30):
> display({pdome,pV});
```

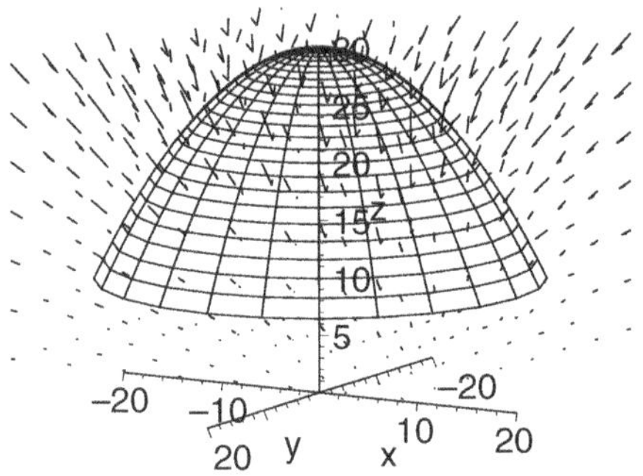

In example 6.11, we parametrized the silo dome as $\vec{R}(r,\theta) = \left(r\cos\theta, r\sin\theta, 30 - \frac{r^2}{20}\right)$ and computed the normal to be:

```
> N;
```

$$\left[\frac{1}{10}\,r^2\cos(\theta),\ \frac{1}{10}\,r^2\sin(\theta),\ r\right]$$

Since $N_z = r > 0$, the normal points upward and we must reverse it to point downward:

```
> N:=-N;
```

$$N := \left[-\frac{1}{10}\,r^2\cos(\theta),\ -\frac{1}{10}\,r^2\sin(\theta),\ -r\right]$$

When $\vec{V}$ is restricted to the silo dome, the composition is

```
> VR:=V &@ R &@ <r,theta>;
```

$$VR := \left[r\sin(\theta)\,(30 - \frac{r^2}{20}),\ (30 - \frac{r^2}{20})\,r\cos(\theta),\ -(30 - \frac{r^2}{20})^2\right]$$

The integral of $\vec{V}$ over the silo dome is $\iint_R \vec{V}\cdot d\vec{S} = \int_0^{2\pi}\int_0^{20} \vec{V}(\vec{R}(\theta,\phi))\cdot\vec{N}\,dr\,d\theta$. Using *Maple*, the integral is

```
> Muint(VR &. N, r=0..20, theta=0..2*Pi); value(%);
```

$$\int_0^{2\pi}\int_0^{20} \frac{1}{400}\,(-600 + r^2)\,r\,(4\,r^2\sin(\theta)\cos(\theta) - 600 + r^2)\,dr\,d\theta$$

$$\frac{520000\,\pi}{3}$$

The surface integral of a vector field can also be written as:

$$\iint_R \vec{F} \cdot d\vec{S} = \iint_R F_1(x,y,z)\,dy\,dz + F_2(x,y,z)\,dz\,dx + F_3(x,y,z)\,dx\,dy$$

In this form, the integral is computed (by hand) by replacing the coordinates by their values on the curve and the products of differentials by the appropriate Jacobian determinants. The result is integrated with respect to the parameters. However, on the computer, it is still easier to compute the dot product of the vector field and the normal.

Example: 6.15. Compute the surface integral $\iint_S -x^2y\,dy\,dz + y^2x\,dz\,dx + z^3\,dx\,dy$ over the spiral ramp $\vec{R}(r,\theta) = (r\cos\theta, r\sin\theta, \theta)$ for $0 \le r \le 5$ and $0 \le \theta \le 3\pi$ and with the normal pointing upward.

SOLUTION: We input the surface and compute the tangent vectors and the normal vector:

```
> R:=MF(<r,theta>, <r*cos(theta)| r*sin(theta)| theta>);
```

$$R := [(r,\theta) \to r\cos(\theta),\ (r,\theta) \to r\sin(\theta),\ (r,\theta) \to \theta]$$

```
> Rr:=D[1](R); Rtheta:=D[2](R);
```

$$Rr := [(r,\theta) \to \cos(\theta),\ (r,\theta) \to \sin(\theta),\ 0]$$

$$Rtheta := [(r,\theta) \to -r\sin(\theta),\ (r,\theta) \to r\cos(\theta),\ 1]$$

```
> N:=(Rr &@ <r,theta>) &x (Rtheta &@ <r,theta>);
```

$$N := [\sin(\theta),\ -\cos(\theta),\ r]$$

Since $N_z = r > 0$, the normal points upward and we do not need to reverse the normal. Next we look at the coefficients of $dy\,dz$, $dz\,dx$ and $dx\,dy$ to identify the vector field and evaluate the vector field on the curve:

```
> F:=MF(<x,y,z>, <-x^2*y| y^2*x| z^3>);
```

$$F := [(x,y,z) \to -x^2\,y,\ (x,y,z) \to y^2\,x,\ (x,y,z) \to z^3]$$

```
> FR:=F &@ R &@ <r,theta>;
```

$$FR := \left[-r^3\cos(\theta)^2\sin(\theta),\ r^3\sin(\theta)^2\cos(\theta),\ \theta^3\right]$$

Finally, we compute the surface integral:

```
> Muint(FR &.  N, r=0..5, theta=0..3*Pi); value(%);
```

$$\int_0^{3\pi}\int_0^5 r\,(\theta^3 - 2\,r^2\cos(\theta)^2 + 2\,r^2\cos(\theta)^4)\,dr\,d\theta$$

$$\frac{2025}{8}\,\pi^4 - \frac{1875}{16}\,\pi$$

6.2.6 Flux and Expansion

Table C.3 in Appendix C, shows the standard applications of surface integrals of vector functions. As examples, we consider the flux of a fluid through an open surface and the expansion of a fluid through a closed surface. These applications may also be applied to an electric or magnetic field.

Flux If a fluid flows through a surface $\vec{R}(u,v)$, then the volume of fluid that flows through the surface per unit time is called the flux of the fluid through the surface and is computed as the surface integral of the normal component of the velocity field over the surface. The flux will be positive if the fluid flows through the surface in the direction of the normal vector. In particular, if the fluid velocity field is $\vec{V}_f$, then the flux is:

$$\mathcal{F}lux = \iint\limits_R \vec{V}_f \cdot \hat{N}\, dS = \iint\limits_R \vec{V}_f \cdot d\vec{S}.$$

Example: 6.16. Suppose a fluid is moving with the velocity field $\vec{V}_f = (x-y, x+y, 2z)$. Find the flux of the fluid outward through the piece of the cylinder $x^2 + y^2 = 4$ for $0 \le z \le 3$. Then plot the fluid velocity and the cylinder in the same plot.

SOLUTION: The surface may be parametrized as

```
> R:=MF(<theta,z>, <2*cos(theta)| 2*sin(theta)| z>);
```

$$R := [(\theta,\ z) \to 2\cos(\theta),\ (\theta,\ z) \to 2\sin(\theta),\ (\theta,\ z) \to z]$$

So the tangent and normal vectors are

```
> Rtheta:=D[1](R); Rz:=D[2](R);
```

$$Rtheta := [(\theta,\ z) \to -2\sin(\theta),\ (\theta,\ z) \to 2\cos(\theta),\ 0]$$

$$Rz := [0,\ 0,\ 1]$$

```
> N:=Cross(Rtheta &@ <theta,z>, Rz &@ <theta,z>);
```

$$N := [2\cos(\theta),\ 2\sin(\theta),\ 0]$$

Examining the sign of the x- and y-components of the normal, we verify that the normal points outward. We now input the velocity field and restrict it to the surface:

```
> Vf:=MF(<x,y,z>, <x-y| x+y| 2*z>);
```

$$Vf := [(x,\ y,\ z) \to x - y,\ (x,\ y,\ z) \to x + y,\ (x,\ y,\ z) \to 2\,z]$$

```
> VfR:=Vf &@ R &@ <theta,z>;
```

$$VfR := [2\cos(\theta) - 2\sin(\theta),\ 2\cos(\theta) + 2\sin(\theta),\ 2\,z]$$

Then the flux is

```
> Muint( VfR &.  N, theta=0..2*Pi, z=0..3 );  Flux_cyl:=value(%);
```

$$\int_0^3 \int_0^{2\pi} 4\, d\theta\, dz$$

$$Flux\_cyl := 24\,\pi$$

We plot the fluid velocity and the cylinder and then display them together:

```
> pVf:=fieldplot3d(Vf &@ <x,y,z>, x=-3..3, y=-3..3, z=0..3):
> pR:=plot3d(R &@ <theta|z>, theta=0..2*Pi, z=0..3):
> display({pVf, pR}, axes=normal, orientation=[35,75]);
```

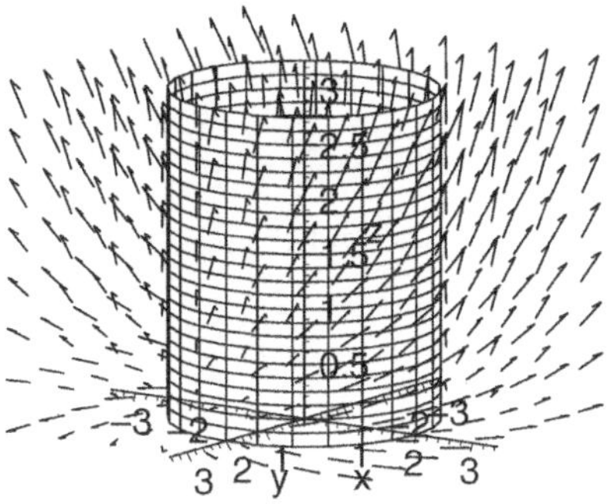

Expansion If the fluid flows through a closed surface, then the flux through the surface (with the outward normal) is called the expansion of the fluid out of the solid region enclosed by the surface.

Example: 6.17. Suppose a fluid is moving with the velocity field $\vec{V}_f = (x - y, x + y, 2z)$. Find the expansion of the fluid out of the solid cylinder $x^2 + y^2 \le 4$ for $0 \le z \le 3$.

SOLUTION: The flux of the fluid through the curved surface of the cylinder was found in the previous example. It remains to compute the flux through the top and bottom surfaces with outward normals. Both the top and bottom surfaces may be parametrized as

```
> R:=MF(<r,theta>, <r*cos(theta)| r*sin(theta)| z0>);
```

$$R := [(r, \theta) \to r\cos(\theta), (r, \theta) \to r\sin(\theta), (r, \theta) \to z0]$$

where z0 is 3 for the top and 0 for the bottom. In both cases the tangent and normal vectors are

```
> Rr:=D[1](R); Rtheta:=D[2](R);
```

$$Rr := [(r, \theta) \to \cos(\theta), (r, \theta) \to \sin(\theta), 0]$$

$$Rtheta := [(r, \theta) \to -r\sin(\theta), (r, \theta) \to r\cos(\theta), 0]$$

```
> N:=Cross(Rr &@ <r,theta>,Rtheta &@ <r,theta>);
```

$$N := [0, 0, r]$$

Examining the sign of the z-component of the normal, we see that this normal points upward. So we need to reverse the normal for the bottom:

```
> N2:=-N;
```

$$N2 := [0, 0, -r]$$

We now restrict the vector field to the surface:

```
> VfR:=Vf &@ R &@ <r,theta>;
```

$$VfR := [r\cos(\theta) - r\sin(\theta), r\cos(\theta) + r\sin(\theta), 2\,z0]$$

Then the flux through the top is

```
> z0:=3; Muint( VfR &.  N, r=0..2, theta=0..2*Pi );  Flux_top:=
> value(%);
```

$$z0 := 3$$

$$\int_0^{2\pi} \int_0^2 6\, r\, dr\, d\theta$$

$$Flux\_top := 24\,\pi$$

and the flux through the bottom is

```
> z0:=0; Muint( VfR &.  N2, r=0..2, theta=0..2*Pi );  Flux_bot:=
> value(%);
```

$$z0 := 0$$

$$\int_0^{2\pi} \int_0^2 0\, dr\, d\theta$$

$$Flux\_bot := 0$$

Adding the fluxes, we have that the net expansion is

```
> Expansion:=Flux_cyl + Flux_top + Flux_bot;
```

$$Expansion := 48\,\pi$$

Since the net expansion is positive, we see that there is a net flow of fluid out of the surface.

6.3 Exercises

- Do Project: 10.9 and 10.12.

1. If a wire has the shape of the spiral $\vec{r}(t) = \big(t\cos(t), t\sin(t)\big)$ for $0 \le t \le 2\pi$ and has linear density $\rho(t) = t$, compute the mass and center of mass of the wire.

2. If a wire has the shape of the astroid $\vec{r}(t) = \big(\cos^3(t), \sin^3(t)\big)$ for $0 \le t \le 2\pi$ (See Chapter 2 exercise 2.) and has linear density $\rho(x, y) = x^2 + y^2$, compute the mass and center of mass of the wire.

3. Compute the integral which represents the work done by the force field $\vec{F}(x, y) = (x, y+2)$ in moving an object along an arch of the cycloid $\vec{r}(t) = (t - \sin t, 1 - \cos t)$ for $0 \le t \le 2\pi$.

4. Compute $\oint_C (-y^3\, dx + x^3\, dy)$ once counterclockwise around the circle $x^2 + y^2 = 9$.

5. Repeat example 6.7 but for the velocity field: $\vec{V}_f = \left(\dfrac{-y}{x^2+y^2}, \dfrac{x}{x^2+y^2}\right)$. For each family of circles, plot the circulation as a function of the radius a and discuss the dependence on a.

   NOTE: Before computing the integral for the second family of circles, you will need to impose the assumption $0 < a < 2$ or separately $2 < a$ by using *Maple*'s `assume` command. (Otherwise, *Maple* is unable to do the integral.) You do this by executing: `assume(0<a,a<2);` or `assume(2<a);`

6. The surface of a sphere of radius 2 may be parametrized as

$$\vec{R}(\theta,\phi) = (2\sin(\phi)\cos(\theta), 2\sin(\phi)\sin(\theta), 2\cos(\phi))\ .$$

Plot the sphere and compute the coordinate tangent vectors and the normal vector. Does the normal point radially inward or outward?

7. What percent of the earth's surface is above the arctic circle at 66.5° north latitude? (In spherical coordinates that's at $\phi = 23.5°$.)

8. Find the area of the part of the parametric surface $\vec{R}(u,w) = (u+w, u-w, w)$ that lies over the triangular shadow region in the xy-plane with vertices (0, 0), (4, 2), and (0, 2).

9. Evaluate the surface integral $\iint_S y\,dS$, where the surface S is the part of the plane $3x + 2y + z = 6$ that lies in the first octant.

10. Find the mass, center of mass and moment of inertia about the z-axis of the spiral ramp of example 6.8 for $0 \le r \le 9$ and $0 \le \theta \le 8\pi$ if the mass density is given by $\rho = 2z$.

NOTE: The center of mass may not be on the z-axis.

11. Consider a thin funnel whose conically-shaped surface S is given by $z = \sqrt{x^2+y^2}$ for $1 \le z \le 4$. Use *cylindrical* coordinates to parametrize S. The projection of the funnel surface onto the xy-plane is a ring. Accordingly, what are the ranges for r and θ? Compute the mass and center of mass of the funnel, given that its *variable* density is $\rho = 10 - z$.

12. Consider the sphere of radius 2 centered at the origin. Compute the integral of the vector field $\vec{F}(x,y,z) = (yz, xz, z^2)$ over the eighth of the sphere in the first octant with normal pointing away from the origin.

13. Suppose the velocity field of a fluid is the radial vector field $\vec{V}_f = (x, y, z)$. Plot the velocity field and compute the expansion of the fluid out of the two families of spheres $x^2+y^2+z^2 = a^2$ and $x^2+y^2+(z-2)^2 = a^2$. For each family of spheres, relate the expansion to the volume of the sphere. This is not a coincidence and will be explained in subsection 8.5.2 using Gauss' Theorem.

14. Repeat exercise 13 but for the velocity field,
$\vec{V}_f = \left(\frac{x}{(x^2+y^2+z^2)^{3/2}}, \frac{y}{(x^2+y^2+z^2)^{3/2}}, \frac{z}{(x^2+y^2+z^2)^{3/2}}\right)$. For both families of spheres, plot the expansion as a function of the radius a and discuss the dependence on a.

NOTE: Before computing the integral for the second family of spheres, you will need to `assume` $0 < a < 2$ or $2 < a$. See exercise 5.

15. In physics, the integral form of Gauss' Law for electrostatics is $\iint_S \vec{E}\cdot\vec{dS} = 4\pi Q$ where $\vec{E}$ is the electric field and Q is the total electric charge inside the closed surface S. Compute the total charge inside the cylinder $x^2+y^2 \le a^2$ for $0 \le z \le h$ for each of the following electric fields. (k is a constant.)

(a) $\vec{E} = (kx(x^2+y^2), ky(x^2+y^2), 0)$

(b) $\vec{E} = (kx, ky, 0)$

(c) $\vec{E} = (\frac{kx}{x^2+y^2}, \frac{ky}{x^2+y^2}, 0)$

Which one of these fields represents the electric field of a line of charge along the z-axis? Its total charge is proportional to the length h and independent of the radius a of the cylinder. Which one of these fields represents the electric field of a uniform charge distribution? Its total charge is proportional to the volume of the cylinder.

16. An electric charge distribution along a curve is specified by giving the linear charge density λ which is measured in units of charge per unit length. Assume there is a uniform charge density along the z-axis with constant charge density λ. Thus the charge in a length L is $Q = L\lambda$. This produces an electric field $\vec{E} = (\frac{2\lambda x}{x^2+y^2}, \frac{2\lambda y}{x^2+y^2}, 0)$.

    (a) Compute the flux of the electric field through a cylinder of radius r and length L centered on the z-axis.

    (b) Compute the flux of the electric field through a cylinder of radius $r < a$ and length L centered on the line $x = a$ and $y = 0$.
    NOTE: Before computing the integral, `assume` $0 < r < a$. See exercise 5.

    (c) Compute the flux of the electric field through a cylinder of radius $r > a$ and length L centered on the line $x = a$ and $y = 0$.
    NOTE: This time `assume` $0 < a < r$.

    (d) Discuss the results by relating the flux to the amount of charge inside the cylinder. (See exercise 15 and lab 9.9.)

17. Assume there is a uniform electric current I moving up the z-axis. This produces the magnetic field $\vec{B} = \frac{2I}{x^2+y^2}(-y, x, 0)$.

    (a) Compute the circulation of the magnetic field counterclockwise around the circle $x^2 + y^2 = r^2$ in the xy-plane.

    (b) Compute the circulation of the magnetic field counterclockwise around the circle $(x-a)^2 + y^2 = r^2$ in the xy-plane if $r < a$.
    NOTE: Before computing the integral, `assume` $0 < r < a$. See exercise 5.

    (c) Compute the circulation of the magnetic field counterclockwise around the circle $(x-a)^2 + y^2 = r^2$ in the xy-plane if $r > a$.
    NOTE: This time `assume` $0 < a < r$.

    (d) Discuss the results by relating the circulation to the current passing through the circle. (See lab 9.10.)

Chapter 7

Vector Differential Operators

7.1 The Del Operator and the Gradient

In rectangular coordinates, it is useful to introduce the "del"-operator given by

$$\vec{\nabla} = \left(\frac{\partial}{\partial x}, \frac{\partial}{\partial y}, \frac{\partial}{\partial z} \right) = \hat{\imath}\frac{\partial}{\partial x} + \hat{\jmath}\frac{\partial}{\partial y} + \hat{k}\frac{\partial}{\partial z} \, .$$

Thus $\vec{\nabla}$ is the vector whose components are the partial derivative operators. Throughout this chapter, we will use the $\vec{\nabla}$ operator to construct other differential operators. However,
BEWARE: The $\vec{\nabla}$ operator only makes sense in rectangular coordinates.

For example, when the $\vec{\nabla}$ operator is applied to a scalar field (i.e. a function) f, it produces a vector field whose entries are the partial derivatives of f. In subsection 3.1.4 this was defined to be the gradient of the function:

$$\vec{\nabla} f = \text{grad} f = \left(\frac{\partial f}{\partial x}, \frac{\partial f}{\partial y}, \frac{\partial f}{\partial z} \right)$$

It may be computed using the command `grad` in the `linalg` package which acts on expressions or using the command `Gradient` in the `VecCalc` package or its alias `Grad` which acts on arrow-defined functions. Examples appear in subsection 3.1.4. The applications, interpretation and plots of the gradient are discussed in subsections 3.2.4, 3.2.5, 3.2.6 and 3.2.7.

7.2 Divergence

7.2.1 Computation

The divergenceof a vector field $\vec{F} = (F_1, F_2, F_3)$ is defined by

$$\text{div}\vec{F} = \vec{\nabla} \cdot \vec{F} = \frac{\partial F_1}{\partial x} + \frac{\partial F_2}{\partial y} + \frac{\partial F_3}{\partial z}$$

where we have interpreted the formula as the dot product of the $\vec{\nabla}$ operator and the vector field $\vec{F}$ except that multiplication has been replaced by differentiation. The result is a scalar field (or function).

The interpretation of the divergence is discussed in lab 9.11. Basically, the divergence at a point P measures the expansion of the vector field out of a small sphere at P (or out of a circle in $\mathbb{R}^2$).

In *Maple*, you may compute the divergence by using the `linalg` command `diverge` for a vector or list of expressions or by using the `VecCalc` command `Div` for a Vector or list of arrow-defined functions.

Example: 7.1. Plot each of the following 2-dimensional vector fields for $-2 \le x \le 2$ and $-2 \le y \le 2$ along with a circle centered at $(1,1)$ of radius .5.

a) $\vec{r} = (x, y)$

b) $\vec{\omega} = (y, -x)$

From the plot, is the divergence positive, negative or zero? (It is positive, if more or bigger vectors come out of the circle than go in; negative, if more go in than out.) Then compute the divergence.

SOLUTION: We enter each vector field using `MF`, plot it using `fieldplot` and `display` it with the circle produced using the `circle` command from the `plottools` package. Then we compute the divergence using `Div`.

a) For $\vec{r} = (x, y)$ we have

```
> r:=MF([x,y], <x,y>);
```

$$r := \begin{bmatrix} (x, y) \to x \\ (x, y) \to y \end{bmatrix}$$

```
> fp:=fieldplot(r &@ <x,y>, x=-2..2, y=-2..2):
> with(plottools);
> display(fp, circle([1,1],.5), scaling=constrained);
```

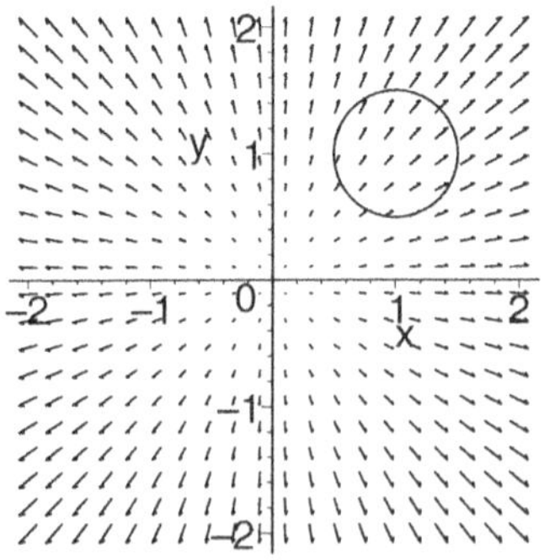

CAUTION: The vectors produced by `fieldplot` may not have the correct lengths. They have been scaled to look good in the plot. However, the directions and relative lengths are correct.

Notice that bigger vectors come out of the circle. So the divergence should be positive. It is:

```
> div_r:=Div(r);
```

$$div\_r := (x,\, y) \to 2$$

b) For $\vec{\omega} = (y, -x)$ we have

```
> omega:=MF([x,y], <y,-x>):
> fp:=fieldplot(omega &@ <x,y>, x=-2..2, y=-2..2):
> display(fp, circle([1,1],.5), scaling=constrained);
```

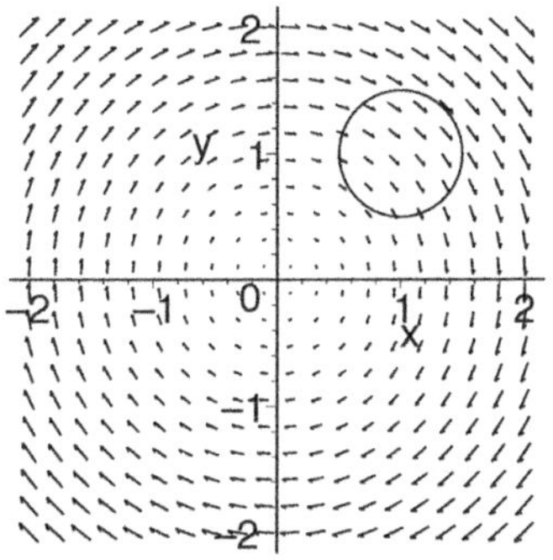

Notice that the same size and number of vectors go in as out of the circle. So the divergence should be zero. It is:

```
> div_omega:=Div(omega);
```

$$div\_omega := (x,\, y) \to 0$$

Example: 7.2. Plot the 3-dimensional vector field $\vec{F} = (0, -2y, 0)$ for $-2 \le x \le 2$, $-2 \le y \le 2$ and $-2 \le z \le 2$ along with a sphere centered at $(1, 1, 1)$ of radius 1. From the plot, is the divergence positive, negative or zero? (It is positive, if more or bigger vectors come out of the sphere than go in; negative, if more go in than out.) Then compute the divergence.

SOLUTION: We first enter the vector field using `MF`, plot it using `fieldplot3d` and `display` it with the sphere produced using the `sphere` command from the `plottools` package. Then we compute the divergence using `Div`:

```
> F:=MF([x,y,z], <0| -2*y| 0>);
```

$$F := [(x,\, y,\, z) \to 0,\, (x,\, y,\, z) \to -2\,y,\, (x,\, y,\, z) \to 0]$$

```
> fp:=fieldplot3d(F &@ <x,y,z>, x=-2..2, y=-2..2, z=-2..2):
> display(fp,sphere([1,1,1],1), scaling=constrained,
> orientation=[20,70]);
```

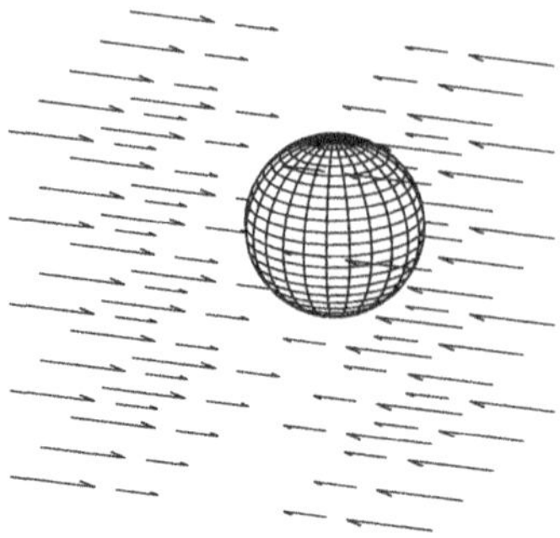

CAUTION: The vectors produced by `fieldplot3d` may not have the correct lengths. They have been scaled to look good in the plot. However, the directions and relative lengths are correct.

Notice that bigger vectors go into the sphere. So the divergence should be negative. It is:

```
> div_F:=Div(F);
```

$$div\_F := (x, y, z) \to -2$$

7.2.2 Applications

Example: 7.3. In physics, the differential form of Gauss' Law for electrostatics is $\vec{\nabla} \cdot \vec{E} = 4\pi\rho_c$ where $\vec{E}$ is the electric field and ρ_c is the electric charge density. Compute the charge density for each of the following electric fields. (k is a constant.) See exercise 6.15.

a) $\vec{E} = (kx(x^2 + y^2), ky(x^2 + y^2), 0)$

b) $\vec{E} = (kx, ky, 0)$

c) $\vec{E} = (\dfrac{kx}{x^2 + y^2}, \dfrac{ky}{x^2 + y^2}, 0)$

For each field, compute the total charge inside the cylinder $x^2 + y^2 \le a^2$ for $0 \le z \le h$.
One of these fields represents the electric field of a line of charge along the z-axis. Which one?
One of these fields represents the electric field of a uniform charge distribution. Which one?

SOLUTION: (To save space, we omit some output.)

a) For $\vec{E} = (kx(x^2 + y^2), ky(x^2 + y^2), 0)$, we compute

```
> E:=MF([x,y,z], <k*x * (x^2+y^2), k*y * (x^2+y^2), 0>):
> rho:=simplify(1/(4*Pi)* Div(E) &@ <x,y,z>);
```

$$\rho := \frac{k\,(x^2 + y^2)}{\pi}$$

Since the density is non-zero and non-constant, this electric field is neither a line of charge along the z-axis nor a uniform distribution. In cylindrical coordinates, the density is

```
> rho:=k*r^2/Pi:
```

So the total charge is

```
> Q:=Muint(rho*r, r=0..a, theta=0..2*Pi, z=0..h); Q:=value(%);
```

$$Q := \int_0^h \int_0^{2\pi} \int_0^a \frac{k\, r^3}{\pi}\, dr\, d\theta\, dz$$

$$Q := \frac{k\, a^4\, h}{2}$$

This should agree with your result from exercise 6.15(a). b) For $\vec{E} = (kx, ky, 0)$, we compute

```
> E:=MF([x,y,z], [k*x, k*y, 0]):
```

```
> rho:=1/(4*Pi)* Div(E)(x,y,z);
```

$$\rho := \frac{k}{2\,\pi}$$

Since this is constant, $\vec{E}$ is the electric field for a uniform charge distribution. So the total charge is

```
> Q:=Muint(rho*r, r=0..a, theta=0..2*Pi, z=0..h); Q:=value(%);
```

$$Q := \int_0^h \int_0^{2\pi} \int_0^a \frac{k\, r}{2\,\pi}\, dr\, d\theta\, dz$$

$$Q := \frac{k\, a^2\, h}{2}$$

This should agree with your result from exercise 6.15(b). c) For $\vec{E} = (\frac{kx}{x^2+y^2}, \frac{ky}{x^2+y^2}, 0)$, we compute

```
> E:=MF([x,y,z], <k*x / (x^2+y^2), k*y / (x^2+y^2), 0>):
```

```
> rho:=1/(4*Pi)* Div(E)(x,y,z);
```

$$\rho := 0$$

Since the density is zero, there does not appear to be any charge anywhere. However, this conclusion is based on Gauss' equation which only holds at points where the electric field is well-defined. Since the electric field is undefined (infinite) on the z-axis, Gauss' equation fails and there may still be charge along the z-axis. In fact, you should have found in example 6.15(c) that there is a uniform charge density along the z-axis.

7.3 Curl

7.3.1 Computation

In $\mathbb{R}^3$, the curlof a vector field $\vec{F} = (F_1, F_2, F_3)$ is defined by

$$\text{curl}\vec{F} = \vec{\nabla} \times \vec{F} = \begin{vmatrix} \hat{\imath} & \hat{\jmath} & \hat{k} \\ \dfrac{\partial}{\partial x} & \dfrac{\partial}{\partial y} & \dfrac{\partial}{\partial z} \\ F_1 & F_2 & F_3 \end{vmatrix} = \left(\frac{\partial F_3}{\partial x_2} - \frac{\partial F_2}{\partial x_3}, \frac{\partial F_1}{\partial x_3} - \frac{\partial F_3}{\partial x_1}, \frac{\partial F_2}{\partial x_1} - \frac{\partial F_1}{\partial x_2} \right)$$

where we have interpreted the formula as the cross product of the $\vec{\nabla}$ operator and the vector field $\vec{F}$ except that multiplication has been replaced by differentiation. The result is a 3-dimensional vector field.

The interpretation of the curl is discussed in lab 9.12. Basically, the direction of the curl specifies the axis and direction of rotation (by the right hand rule) while the magnitude of the curl specifies the rate of rotation.

In *Maple*, you may compute the curl by using the `linalg` command `curl` for a vector or list of expressions or by using the `VecCalc` command `Curl` for a Vector or list of arrow-defined functions.

Example: 7.4. Each of the following vector fields was plotted as a 2-dimensional vector field in example 7.1. Now regard each as a 3-dimensional vector field by appending 0 as the z-component.

a) $\vec{r} = (x, y, 0)$

b) $\vec{\omega} = (y, -x, 0)$

By examining the circulation around the circle, is the z component of the curl positive (counterclockwise), negative (clockwise) or zero? Now compute the curl. Notice that each curl ends up pointing in the z-direction. Is the z component positive, negative or zero?

Solution: After examining the plot, we re-enter the vector field using `MF`, and compute the curl using `Curl`.

a) For $\vec{r} = (x, y, 0)$ we examine the plot and see that the vectors circulate around the circle clockwise as much as counterclockwise. So we expect the z component of the curl is zero. We compute:

```
> r:=MF([x,y,z], <x| y| 0>):
> curl_r:=Curl(r);
```

$$curl\_r := [(x, y, z) \to 0, (x, y, z) \to 0, (x, y, z) \to 0]$$

b) For $\vec{\omega} = (y, -x, 0)$ we examine the plot and see that bigger vectors circulate clockwise around the circle. So we expect the z component of the curl is negative. We compute:

```
> omega:=MF([x,y,z], <y| -x| 0>):
> curl_omega:=Curl(omega);
```

$$curl\_omega := [(x, y, z) \to 0, (x, y, z) \to 0, (x, y, z) \to -2]$$

Example: 7.5. Compute the curl of the vector field $\vec{u} = (-z, 0, x)$ and convert the curl to spherical coordinates. Then plot $\vec{u}$ and orient the plot according to the direction of $\vec{\nabla} \times \vec{u}$. What do you notice?

SOLUTION: We enter $\vec{u}$, compute the curl and convert to spherical coordinates:

```
> u:=MF([x,y,z], <-z|0|x>):
> curl_u:=Curl(u);
```

$$curl\_u := [(x, y, z) \to 0, (x, y, z) \to -2, (x, y, z) \to 0]$$

```
> r2s(curl_u &@ <x,y,z>);
```

$$\left[2, -\frac{\pi}{2}, \frac{\pi}{2}\right]$$

So the curl points in the direction $\theta = -\frac{\pi}{2}$ rad $= -90°$ and $\phi = \frac{\pi}{2}$ rad $= 90°$. We now plot $\vec{u}$, taking the orientation to be the direction of the curl:

```
> fieldplot3d(u &@ <x,y,z>, x=-2..2, y=-2..2, z=-2..2,
> orientation=[-90,90], arrows='3-D');
```

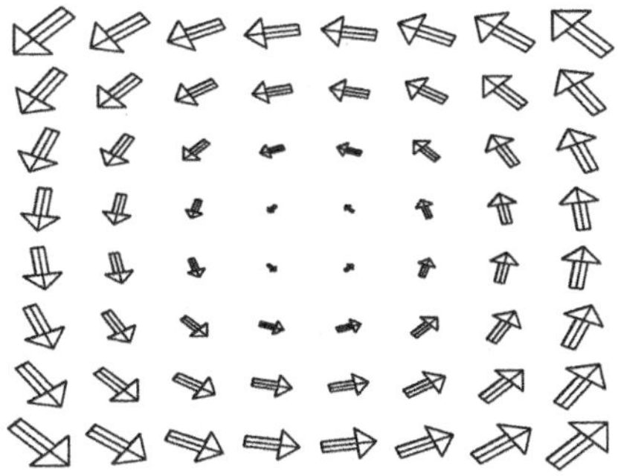

Notice that the vector field rotates counterclockwise around the direction of the curl. (Try dragging the plot with your mouse.)

7.3.2 Applications

Example: 7.6. In physics, the differential form of Ampere's Law for magnetostatics is $\vec{\nabla} \times \vec{B} = 4\pi\vec{J}$ where $\vec{B}$ is the magnetic field and $\vec{J}$ is the electric current density. (A current has the units of charge per unit time, while a current density has the units of charge per unit time per unit area. Thus if you integrate a current density over a surface, you get the total current passing through that surface.) Compute the current density for the magnetic field $\vec{B} = (x^3z, y^3z, 0)$.

SOLUTION: We enter the magnetic field and compute its curl:

```
> B:=MF([x,y,z], <x^3*z| y^3*z| 0>);
```

$$B := [(x, y, z) \to x^3 z, (x, y, z) \to y^3 z, (x, y, z) \to 0]$$

```
> curlB:=Curl(B);
```

$$curlB := [(x, y, z) \to -y^3, (x, y, z) \to x^3, (x, y, z) \to 0]$$

Dividing by 4π gives the current density:

```
> J:=(1/(4*Pi)*curlB &@ <x,y,z>);
```

$$J := \left[-\frac{y^3}{4\,\pi}, \frac{x^3}{4\,\pi}, 0\right]$$

Notice that just because a vector field lies in the xy-plane does not mean that its curl must be perpendicular to the xy-plane.

7.4 Higher Order Differential Operators and Identities

In this section, we will investigate the second and higher order differential operators which may be constructed from the gradient, divergence and curl, and the identities satisfied by some of these.

7.4.1 Laplacian of a Scalar

The divergence of the gradient of a scalar f is also called the Laplacian of f and is simply the sum of the second partial derivatives of f with respect to each variable:

$$\vec{\nabla} \cdot \vec{\nabla} f = \vec{\nabla}^2 f = \text{Lap}(f) = \frac{\partial^2 f}{\partial x^2} + \frac{\partial^2 f}{\partial y^2} + \frac{\partial^2 f}{\partial z^2}$$

A function satisfying Laplace's equation, $\vec{\nabla}^2 f = 0$, is called harmonic.

In *Maple*, you may compute the Laplacian by using the `linalg` command `laplacian` for expressions or by using the `VecCalc` command `Lapliaian` or its alias `Lap` for arrow-defined functions.

Example: 7.7. Compute the Laplacian of each of the following functions. Which is harmonic?

a) $f = ax^2 + by^2 + cz^2 + 2pyz + 2qxz + 2rxy$ where a, b, c, p, q and r are constants.

b) $g = e^x \cos(y)$

SOLUTION: For each vector field, we enter the function using `MF` and compute the Laplacian using `Lap`:

a) For $f = ax^2 + by^2 + cz^2 + 2pyz + 2qxz + 2rxy$ we compute

```
> f:=MF([x,y,z], a*x^2+b*y^2+c*z^2+2*p*y*z+2*q*x*z+2*r*x*y);
```

$$f := (x, y, z) \to a\,x^2 + b\,y^2 + c\,z^2 + 2\,p\,y\,z + 2\,q\,x\,z + 2\,r\,x\,y$$

```
> Lf:=Lap(f);
```

$$Lf := (x, y, z) \to 2\,a + 2\,b + 2\,c$$

b) For $g = e^x \cos(y)$ we compute

```
> g:=MF([x,y], exp(x)*cos(y));
```

$$g := (x, y) \to e^x \cos(y)$$

```
> Lg:=Lap(g);
```

$$Lg := (x,\ y) \to 0$$

So g is harmonic.

7.4.2 Laplacian of a Vector

The divergence of the gradient of a vector $\vec{F} = (F_1, F_2, F_3)$ is called the Laplacian of $\vec{F}$ and is simply the Laplacian of each component of $\vec{F}$:

$$\vec{\nabla} \cdot \vec{\nabla} \vec{F} = \vec{\nabla}^2 \vec{F} = \text{Lap}(\vec{F}) = \frac{\partial^2 \vec{F}}{\partial x^2} + \frac{\partial^2 \vec{F}}{\partial y^2} + \frac{\partial^2 \vec{F}}{\partial z^2} = \left(\vec{\nabla}^2 F_1, \vec{\nabla}^2 F_2, \vec{\nabla}^2 F_3\right)$$

In *Maple*, the `VecCalc` command `Lap` is designed to compute the Laplacian of a Vector of arrow-defined functions. For arrays or lists of expressions, you must `map` the `linalg` command `laplacian` onto the array or list.

Example: 7.8. Compute the Laplacian of the electric field of a point charge:

$$\vec{E} = \frac{\vec{r}}{r^3} = \left(\frac{x}{(x^2+y^2+z^2)^{3/2}}, \frac{y}{(x^2+y^2+z^2)^{3/2}}, \frac{z}{(x^2+y^2+z^2)^{3/2}}\right)$$

Solution: We enter the electric field using `MF` and compute the Laplacian using `Lap`:

```
> E:=MF([x,y,z], < x/(x^2+y^2+z^2)^(3/2), y/(x^2+y^2+z^2)^(3/2),
> z/(x^2+y^2+z^2)^(3/2) >);
```

$$E := \begin{bmatrix} (x,\ y,\ z) \to \dfrac{x}{(x^2+y^2+z^2)^{(3/2)}} \\ (x,\ y,\ z) \to \dfrac{y}{(x^2+y^2+z^2)^{(3/2)}} \\ (x,\ y,\ z) \to \dfrac{z}{(x^2+y^2+z^2)^{(3/2)}} \end{bmatrix}$$

```
> LE:=Lap(E);
```

$$LE := \begin{bmatrix} (x,\ y,\ z) \to 0 \\ (x,\ y,\ z) \to 0 \\ (x,\ y,\ z) \to 0 \end{bmatrix}$$

So $\vec{E}$ is a harmonic vector field.

7.4.3 Hessian of a Scalar

The gradient of the gradient of a scalar f is also called the Hessian of f and is simply the matrix of all second partial derivatives of f:

$$\vec{\nabla}\vec{\nabla}f = \text{Hess}(f) = \begin{pmatrix} \dfrac{\partial^2 f}{\partial x \partial x} & \cdots & \dfrac{\partial^2 f}{\partial x \partial z} \\ \vdots & \ddots & \vdots \\ \dfrac{\partial^2 f}{\partial z \partial x} & \cdots & \dfrac{\partial^2 f}{\partial z \partial z} \end{pmatrix}$$

In subsection 4.1.2, we used the leading principal minor determinants of the Hessian to classify the critical points of a function as local maxima or local minima.

In *Maple*, you may compute the Hessian by using the `linalg` command `hessian` for expressions or by using the `VecCalc` command `Hessian` or its alias `Hess` for arrow-defined functions.

Example: 7.9. Compute the Hessian of the function $f = ax^2 + by^2 + cz^2 + 2pyz + 2qxz + 2rxy$.

SOLUTION: We enter the function using `MF` and compute the Hessian using `Hess`:

```
> f:=MF([x,y,z], a*x^2+b*y^2+c*z^2+2*p*y*z+2*q*x*z+2*r*x*y):
> Hf:=Hess(f);
```

$$Hf := \begin{bmatrix} (x, y, z) \to 2a & (x, y, z) \to 2r & (x, y, z) \to 2q \\ (x, y, z) \to 2r & (x, y, z) \to 2b & (x, y, z) \to 2p \\ (x, y, z) \to 2q & (x, y, z) \to 2p & (x, y, z) \to 2c \end{bmatrix}$$

7.4.4 Higher Order Gradients of Scalars

The gradient of a scalar field, f, is the vector, $\vec{\nabla}f$, of first partial derivatives of f, namely $\dfrac{\partial f}{\partial x_i}$.

The second order gradient of f (i.e. the Hessian) is the matrix, $\vec{\nabla}\vec{\nabla}f$, of second partial derivatives of f, namely $\dfrac{\partial^2 f}{\partial x_i \partial x_j}$.

Similarly, the third order gradient of f is the three-dimensional array, $\vec{\nabla}\vec{\nabla}\vec{\nabla}f$, of third partial derivatives of f, namely $\dfrac{\partial^3 f}{\partial x_i \partial x_j \partial x_k}$.

And in general the k-th order gradient of f is the k-dimensional array, $\vec{\nabla}\cdots\vec{\nabla}f$, of k-th partial derivatives of f, namely $\dfrac{\partial^k f}{\partial x_{i_1} \cdots \partial x_{i_k}}$.

These higher dimensional arrays are called tensors and are beyond the scope of this book. However, the higher order partial derivatives have been used in subsection 3.2.3 to construct the Higher Order Taylor Polynomial Approximations.

7.4.5 Curl of a Gradient

The curl of the gradient of a scalar f satisfies the first of two extremely important identities. We first consider some examples.

Example: 7.10. Compute the curl of the gradient of each of the following functions:

a) $f = ax^2 + by^2 + cz^2 + 2pyz + 2qxz + 2rxy$

b) $g = \cos(e^y) - \sin\left(\frac{x}{z}\right)$

SOLUTION: We enter each function using `MF` and compute the gradient and the curl using `Grad` and `Curl`:

a) For $f = ax^2 + by^2 + cz^2 + 2pyz + 2qxz + 2rxy$ we compute

```
> f:=MF([x,y,z], a*x^2+b*y^2+c*z^2+2*p*y*z+2*q*x*z+2*r*x*y):
> grad_f:=Grad(f);
```

$$grad\_f := [(x,\, y,\, z) \to 2\,a\,x + 2\,q\,z + 2\,r\,y,\, (x,\, y,\, z) \to 2\,b\,y + 2\,p\,z + 2\,r\,x,\, (x,\, y,\, z) \to 2\,c\,z + 2\,p\,y + 2\,q\,x]$$

```
> curl_grad_f:=Curl(grad_f);
```

$$curl\_grad\_f := [(x,\, y,\, z) \to 0,\, (x,\, y,\, z) \to 0,\, (x,\, y,\, z) \to 0]$$

b) For $g = \cos(e^y) - \sin\left(\frac{x}{z}\right)$ we compute

```
> g:=MF([x,y,z], cos(exp(y)) - sin(x/z)):
> grad_g:=Grad(g);
```

$$grad\_g := \left[(x,\, y,\, z) \to -\frac{\cos(\frac{x}{z})}{z},\, (x,\, y,\, z) \to -\sin(e^y)\, e^y,\, (x,\, y,\, z) \to \frac{\cos(\frac{x}{z})\, x}{z^2}\right]$$

```
> curl_grad_g:=Curl(grad_g);
```

$$curl\_grad\_g := [(x,\, y,\, z) \to 0,\, (x,\, y,\, z) \to 0,\, (x,\, y,\, z) \to 0]$$

Try several other functions. You'll always get $\vec{0} = (0,0,0)$. From these we conjecture that for any function $f(x,y,z)$, we have the identity

$$\text{curl}(\text{grad} f) = \vec{0} \qquad \text{or} \qquad \vec{\nabla} \times \vec{\nabla} f = \vec{0}$$

You should prove this by hand but you can also prove it using *Maple*: First clear `g` and define `f` to be the undefined function `g`:

```
> g:='g'; f:=MF([x,y,z], g(x,y,z));
```

$$g := g$$

$$f := (x,\, y,\, z) \to g(x,\, y,\, z)$$

Then compute the gradient and curl:

```
> grad_f:=Grad(f);
```

$$grad\_f := [(x, y, z) \to D_1(g)(x, y, z), (x, y, z) \to D_2(g)(x, y, z), (x, y, z) \to D_3(g)(x, y, z)]$$

```
> curl_grad_f:=Curl(grad_f);
```

$$curl\_grad\_f := [(x, y, z) \to 0, (x, y, z) \to 0, (x, y, z) \to 0]$$

You always get $\vec{0} = (0, 0, 0)$.

7.4.6 Divergence of a Curl

The divergence of the curl of a vector $\vec{F}$ satisfies the second extremely important identity.We first consider an example.

Example: 7.11. Compute the divergence of the curl of the vector field $\vec{F} = (\sin(x^y), \cos(y^z), \tan(z^x))$.

SOLUTION: We enter the vector field using `MF` and compute the curl and the divergence using `Curl` and `Div`:

```
> F:=MF([x,y,z], < sin(x^y), cos(y^z), tan(z^x) >):
> curl_F:=Curl(F);
```

$$curl\_F := \begin{bmatrix} (x, y, z) \to \sin(y^z)\, y^z \ln(y) \\ (x, y, z) \to -\dfrac{z^x \ln(z)}{\cos(z^x)^2} \\ (x, y, z) \to -\cos(x^y)\, x^y \ln(x) \end{bmatrix}$$

```
> div_curl_F:=Div(curl_F);
```

$$div\_curl\_F := (x, y, z) \to 0$$

Try several other vector fields. You'll always get 0. From these we conjecture that for any vector field $\vec{F}(x, y, z)$, we have the identity

$$\text{div}(\text{curl}\vec{F}) = 0 \qquad \text{or} \qquad \vec{\nabla} \cdot \vec{\nabla} \times \vec{F} = 0$$

You should prove this by hand but you can also prove it using *Maple*: Start with a general vector field $\vec{F} = (F_1, F_2, F_3)$ and compute the divergence of its curl.

```
> F:=MF([x,y,z], <F1(x,y,z)| F2(x,y,z)| F3(x,y,z)>);
```

$$F := [(x, y, z) \to F1(x, y, z), (x, y, z) \to F2(x, y, z), (x, y, z) \to F3(x, y, z)]$$

```
> div_curl_F:= Div(Curl(F));
```

$$div\_curl\_F := (x, y, z) \to 0$$

You always get 0.

7.4.7 Differential Identities

So far in this chapter we have derived two identities:

1. $\text{curl}(\text{grad} f) = \vec{0}$ or $\vec{\nabla} \times \vec{\nabla} f = \vec{0}$

2. $\text{div}(\text{curl}\vec{F}) = 0$ or $\vec{\nabla} \cdot \vec{\nabla} \times \vec{F} = 0$

There is a third identity relating second derivatives:

3. $\text{curl}(\text{curl}\vec{F}) = \text{grad}(\text{div}\vec{F}) - \text{Lap}(\vec{F})$ or $\vec{\nabla} \times \vec{\nabla} \times \vec{F} = \vec{\nabla}(\vec{\nabla} \cdot \vec{F}) - \vec{\nabla}^2 \vec{F}$

There are many other identities satisfied by the gradient, divergence and curl. The most important of these are the product rules listed here:

4. $\text{grad}(fg) = f\text{grad}g + g\text{grad}f$ or $\vec{\nabla}(fg) = f\vec{\nabla}g + g\vec{\nabla}f$
5. $\text{grad}(\vec{F} \cdot \vec{G}) = (\vec{F} \cdot \text{grad})\vec{G} + (\vec{G} \cdot \text{grad})\vec{F}$ or $\vec{\nabla}(\vec{F} \cdot \vec{G}) = (\vec{F} \cdot \vec{\nabla})\vec{G} + (\vec{G} \cdot \vec{\nabla})\vec{F}$
   where $\vec{F} \cdot \text{grad}$ or $\vec{F} \cdot \vec{\nabla}$ is the directional derivative operator which acts on a vector $\vec{G}$ by differentiating each component of $\vec{G}$.
6. $\text{div}(f\vec{G}) = f\text{div}\vec{G} + \text{grad}f \cdot \vec{G}$ or $\vec{\nabla} \cdot (f\vec{G}) = f\vec{\nabla} \cdot \vec{G} + \vec{\nabla}f \cdot \vec{G}$
7. $\text{div}(\vec{F} \times \vec{G}) = (\text{curl}\vec{F}) \cdot \vec{G} + \vec{F} \cdot (\text{curl}\vec{G})$ or $\vec{\nabla} \cdot (\vec{F} \times \vec{G}) = (\vec{\nabla} \times \vec{F}) \cdot \vec{G} + \vec{F} \cdot (\vec{\nabla} \times \vec{G})$
8. $\text{curl}(f\vec{G}) = \text{grad}f \times \vec{G} + f\text{curl}\vec{G}$ or $\vec{\nabla} \times (f\vec{G}) = \vec{\nabla}f \times \vec{G} + f\vec{\nabla} \times \vec{G}$
9. $\text{curl}(\vec{F} \times \vec{G}) = (\text{curl}\vec{F}) \times \vec{G} - \vec{F} \times (\text{curl}\vec{G})$ or $\vec{\nabla} \times (\vec{F} \times \vec{G}) = (\vec{\nabla} \times \vec{F}) \cdot \vec{G} - \vec{F} \cdot (\vec{\nabla} \times \vec{G})$

These identities will be proved in the exercises. They are all proved by computing the left and right sides and subtracting. You will need to use the `grad`, `diverge` and `curl` commands from the `linalg` package.

7.5 Finding Potentials

In the last section, we proved two important identities:

- "The curl of a gradient is zero." $\vec{\nabla} \times \vec{\nabla} f = \vec{0}$
- "The divergence of a curl is zero." $\vec{\nabla} \cdot \vec{\nabla} \times \vec{A} = 0$

These can be rephrased as the two statements:

- If $\vec{F} = \vec{\nabla} f$, then $\vec{\nabla} \times \vec{F} = 0$.
- If $\vec{G} = \vec{\nabla} \times \vec{A}$, then $\vec{\nabla} \cdot \vec{G} = 0$.

In general, the converses are not always true. They depend on the region on which the vector fields are defined. In particular,

- If $\vec{F}$ is defined in a "nice" region R and $\vec{\nabla} \times \vec{F} = 0$, then $\vec{F} = \vec{\nabla} f$ for some function f defined in R. f is called a scalar potential for $\vec{F}$.

- If $\vec{G}$ is defined in a "nice" region R and $\vec{\nabla}\cdot\vec{G}=0$, then $\vec{G}=\vec{\nabla}\times\vec{A}$ for some vector field $\vec{A}$ defined in R. $\vec{A}$ is called a vector potential for $\vec{G}$.

The meaning of "nice" is different in the two cases. Suffice it to say that if the region R is contractible (it has no holes of any type) then the region is "nice" for both cases.

It remains to explain how the scalar and vector potentials may be found.

7.5.1 Scalar Potentials

Suppose you are given a vector field `F` as a Vector or list of arrow-defined functions as produced by the `MF` command. Then the `VecCalc` command `ScalarPotential` or its alias `SPot` will return a scalar potential for `F`, if it exists and will return `undefined` if there is no scalar potential.

Example: 7.12. Determine if each of the following vector fields has a scalar potential and if it does, find it.

a) $\vec{F}=(yz,xz,xy)$

b) $\vec{u}=(yz,-xz,xy)$

SOLUTION: For each vector field, we enter the function using `MF`, test for existence of a scalar potential using `Curl` and find it (if it exists) using `SPot`.

a) For $\vec{F}=(yz,xz,xy)$, we compute:

```
> F:=MF(<x,y,z>, <y*z|x*z|x*y>):
> Curl(F);
```

$$[(x,\ y,\ z)\rightarrow 0,\ (x,\ y,\ z)\rightarrow 0,\ (x,\ y,\ z)\rightarrow 0]$$

Since the curl is zero and $\vec{F}$ is defined in all of space, there is a scalar potential given by

```
> f:=SPot(F);
```

$$f:=(x,\ y,\ z)\rightarrow y\,z\,x$$

You can check it by computing the gradient:

```
> Grad(f);
```

$$[(x,\ y,\ z)\rightarrow y\,z,\ (x,\ y,\ z)\rightarrow x\,z,\ (x,\ y,\ z)\rightarrow y\,x]$$

which is back to $\vec{F}$.

b) For $\vec{u}=(yz,-xz,xy)$, we compute:

```
> u:=MF(<x,y,z>, <y*z|-x*z|x*y>):
> Curl(u);
```

$$[(x,\ y,\ z)\rightarrow 2\,x,\ (x,\ y,\ z)\rightarrow 0,\ (x,\ y,\ z)\rightarrow -2\,z]$$

Since the curl is not zero, there cannot be a scalar potential:

```
> g:=SPot(u);
```

$$g:=\mathit{undefined}$$

7.5.2 Vector Potentials

Suppose you are given a vector field `G` as a Vector or list of arrow-defined functions as produced by the `MF` command. Then the `VecCalc` command `VectorPotential` or its alias `VPot` will return a vector potential for `G`, if it exists and will return `undefined` if there is no vector potential.

Example: 7.13. Determine if each of the following vector fields has a vector potential and if it does, find it.

a) $\vec{G} = (yz, xz, xy)$

b) $\vec{v} = (xz, xy, yz)$

SOLUTION: For each vector field, we enter the function using `MF`, test for existence of a vector potential using `Div` and find it (if it exists) using `VPot`.

a) For $\vec{G} = (yz, xz, xy)$, we compute:

```
> G:=MF(<x,y,z>, <y*z|x*z|x*y>):
> Div(G);
```

$$(x, y, z) \to 0$$

Since the divergence is zero and $\vec{G}$ is defined in all of space, there is a vector potential given by

```
> A:=VPot(G);
```

$$A := \left[(x, y, z) \to \frac{1}{2}\, x\, z^2 - \frac{1}{2}\, x\, y^2,\ (x, y, z) \to -\frac{1}{2}\, y\, z^2,\ (x, y, z) \to 0\right]$$

You can check it by computing the curl:

```
> Curl(A);
```

$$[(x, y, z) \to y\, z,\ (x, y, z) \to x\, z,\ (x, y, z) \to y\, x]$$

which is back to $\vec{G}$.

b) For $\vec{v} = (xz, xy, yz)$, we compute:

```
> v:=MF([x,y,z], [x*z,x*y,y*z]):
> Div(v);
```

$$(x, y, z) \to z + x + y$$

Since the divergence is not zero, there cannot be a vector potential. *Maple* agrees:

```
> B:=VPot(v);
```

$$B := \mathit{undefined}$$

7.6 Exercises

1. Find the divergence and curl of the vector field $\vec{F}(x, y, z) = (e^{xz}, -2e^{yz}, 3xe^{y})$.

2. Plot each of the following 2-dimensional vector fields for $-1 \le x \le 1$ and $-1 \le y \le 1$ along with a circle centered at $(.75, .75)$ of radius $.25$. From the plot, is the divergence positive,

negative or zero? (It is positive, if more or bigger vectors come out of the circle than go in; negative, if more go in than out.) Then compute the divergence.

(a) $\vec{F} = (x^3, 0)$ (b) $\vec{G} = (0, x^3)$

3. Plot each of the following 3-dimensional vector fields for $-2 \le x \le 2$, $-2 \le y \le 2$ and $-2 \le z \le 2$ along with a sphere centered at $(1,1,1)$ of radius 1. From the plot, is the divergence positive, negative or zero? (It is positive, if more or bigger vectors come out of the sphere than go in; negative, if more go in than out.) Then compute the divergence.

   (a) $\vec{r} = (x, y, z)$ (b) $\vec{u} = (0, 0, z)$ (c) $\vec{v} = (0, z, -y)$

4. Each of the following vector fields was plotted as a 2-dimension al vector field in exercise 2. Now regard each as a 3-dimensional vector field by appending 0 as the z-component.

   (a) $\vec{F} = (x^3, 0, 0)$ (b) $\vec{G} = (0, x^3, 0)$

   By examining the circulation around the circle , is the z component of the curl positive (counterclockwise), negative (clockwise) or zero? Now compute the curl. Notice that each curl ends up pointing in the z-direction. Is the z component positive, negative or zero?

5. For each of the vector fields in exercise 3, rotate the plot to see if there is an axis about which the vector field circulates. (HINT: Orient the plot so one axis points straight at you.) Then compute the curl to see that it points along the axis of rotation.

6. Check that the function $f(x, y, t) = e^{-\left(\frac{x+y}{\sqrt{2}} - ct\right)^2} + 2e^{-\left(\frac{x-y}{\sqrt{2}} + ct\right)^2}$ satisfies the 2-dimensional wave equation $\frac{\partial^2 f}{\partial x^2} + \frac{\partial^2 f}{\partial y^2} - \frac{1}{c^2}\frac{\partial^2 f}{\partial t^2} = 0$. You should define `f` as an expression, take the derivatives using `diff`, evaluate the left side of the wave equation and `simplify` to get zero. Make a movie of the wave (for $c = 2$) by using the commands

```
> f2:=subs(c=2,f);
> animate3d(f2, x=-20..20, y=-20..20, t=-12..12, view=0..2, frames=25,
> orientation=[20,60]);
```

   Then click in the plot and click on the PLAY ARROW on the button bar. Repeat for the functions: $f = e^{-(x-ct)^2} - 2e^{-(y+ct)^2}$ and $f = \sin(x - ct) - \cos(y + ct)$ but plot for $-2\pi \le x \le 2\pi$, $-2\pi \le y \le 2\pi$, $-3 \le z \le 3$, $-\pi \le t \le \pi$.

7. Prove the identities #3 – #9 of subsection 7.4.7. You will need to use the `grad`, `diverge` and `curl` commands from the `linalg` package. See the proofs at the end of subsections 7.4.5 and 7.4.6.

8. Show that $\vec{F}(x, y, z) = (4x^3 + y^2 - 3z, 2xy - 2y - 6yz, -3x - 3y^2 - 8z)$ is a conservative vector field (i.e. $\vec{\nabla} \times \vec{F} = 0$) and find a scalar potential function f such that $\vec{F} = \vec{\nabla} f$.

9. Show that $\vec{F}(x, y, z) = (-12x^2y^3z^3 + 2x^3, -8xz, 6xy^3z^4 - 6x^2z - 3x^4y^2)$ is a solenoidal vector field (i.e. $\vec{\nabla} \cdot \vec{F} = 0$) and find a vector potential $\vec{A}$ such that $\vec{F} = \vec{\nabla} \times \vec{A}$.

10. The electric field of a point charge is $\vec{E} = \left(\frac{x}{(x^2+y^2+z^2)^{3/2}}, \frac{y}{(x^2+y^2+z^2)^{3/2}}, \frac{z}{(x^2+y^2+z^2)^{3/2}} \right)$. Since $\vec{E}$ is not defined at the origin, there may or may not be a scalar potential defined everywhere but the origin even if $\vec{\nabla} \times \vec{E} = 0$. Compute the curl of $\vec{E}$ to see it is zero. Then find the scalar potential and determine where it is undefined. Check the potential by computing its gradient.

11. The magnetic field of an electric current along the z-axis is $\vec{B} = \left(\frac{-y}{x^2+y^2}, \frac{x}{x^2+y^2}, 0 \right)$. Since $\vec{B}$ is not defined on the z-axis, there may or may not be a scalar potential defined everywhere but the z-axis even if $\vec{\nabla} \times \vec{B} = 0$. Compute the curl of $\vec{B}$ to see it is zero. Then find the scalar potential and determine where it is undefined. Check the potential by computing its gradient.

12. The magnetic field of an electric current along the z-axis is $\vec{B} = \left(\frac{-y}{x^2+y^2}, \frac{x}{x^2+y^2}, 0 \right)$. Since $\vec{B}$ is not defined on the z-axis, there may or may not be a vector potential defined everywhere but the z-axis even if $\vec{\nabla} \cdot \vec{B} = 0$. Compute the divergence of $\vec{B}$ to see it is zero. Then find a vector potential and determine where it is undefined. Check the potential by computing its curl.

13. The electric field of a point charge is $\vec{E} = \left(\frac{x}{(x^2+y^2+z^2)^{3/2}}, \frac{y}{(x^2+y^2+z^2)^{3/2}}, \frac{z}{(x^2+y^2+z^2)^{3/2}} \right)$. Since $\vec{E}$ is not defined at the origin, there may or may not be a vector potential defined everywhere but the origin even if $\vec{\nabla} \cdot \vec{E} = 0$. Compute the divergence of $\vec{E}$ to see it is zero. Then find a vector potential and determine where it is undefined. Check the potential by computing its curl.

Chapter 8

Fundamental Theorems of Vector Calculus

8.1 Generalizing the Fundamental Theorem of Calculus

In single variable calculus, the Fundamental Theorem of Calculusshows that the integral and the derivative are essentially inverse operators except for an additive constant. The integral is defined as a limit of Riemann sums, but the Fundamental Theorem of Calculus shows that the integral may also be computed in terms of anti-derivatives. Specifically, the Fundamental Theorem of Calculus may be stated in three forms:

The Fundamental Theorem of Calculus.

$$\frac{d}{dx}\int_a^x f(t)\,dt = f(x) \tag{1}$$

$$\int_a^x \frac{dg(t)}{dt}\,dt = g(x) - g(a) \tag{2}$$

$$\int_a^b f(t)\,dt = F(b) - F(a) \quad \text{where} \quad \frac{dF}{dt} = f(t) \tag{3}$$

The first two forms show that derivatives and integrals are inverse operators except for an additive constant of integration: $-g(a)$. The third form is used to compute integrals in terms of antiderivatives. It is the second form which generalizes to all the theorems of several variable calculus to be discussed in this chapter.

8.2 Fundamental Theorem of Calculus for Curves

The Fundamental Theorem of Calculus for Curves. If $\vec{r}(t)$ is a curve in $\mathbb{R}^n$ traversed from a point $A = \vec{r}(a)$ to a point $B = \vec{r}(b)$ and if f is a differentiable function defined in a neighborhood of the curve, then

$$\int_{A \atop \vec{r}(t)}^{B} \vec{\nabla} f \cdot d\vec{s} = f(B) - f(A)$$

8.2.1 Verification

Example: 8.1. Verify the Fundamental Theorem of Calculus for Curves by computing both sides for the function $f(x,y) = x\cos(y) - y^2\sin(x^2)$ and the curve $\vec{r}(t) = (t\cos(t), t\sin(t))$ for $0 \le t \le 4\pi$.

NOTE: This is not a proof of the theorem because you are not verifying it for a general function and curve.

SOLUTION: We enter the function and the curve using `MF`:

```
> f:=MF([x,y],x*cos(y) - y^2*sin(x^2)):
> r:=MF(t,<t*cos(t)| t*sin(t)>):
```

Then we compute the gradient of the function using `Grad` and evaluate on the curve:

```
> delf:=Grad(f);
```

$$delf := [(x,\, y) \to \cos(y) - 2\,y^2\cos(x^2)\,x,\; (x,\, y) \to -x\sin(y) - 2\,y\sin(x^2)]$$

```
> delfr:=delf &@ (r &@ t);
```

$$delfr := \Big[\cos(t\sin(t)) - 2\,t^3\sin(t)^2\cos(t^2\cos(t)^2)\cos(t),$$
$$-t\cos(t)\sin(t\sin(t)) - 2\,t\sin(t)\sin(t^2\cos(t)^2)\Big]$$

Next, we compute the velocity using `D`:

```
> v:=D(r) &@ t;
```

$$v := [\cos(t) - t\sin(t),\, \sin(t) + t\cos(t)]$$

Finally, we dot the gradient into the velocity and integrate using `Int` and `value` to obtain the left hand side: (This takes *Maple* a relatively long time.)

```
> Int(delfr &. v,t=0..4*Pi); LHS=value(%);
```

$$\int_0^{4\pi} \cos(t\sin(t))\cos(t) - \cos(t\sin(t))\,t\sin(t) - 2\,t^3\,\%1\cos(t)^2 + 2\,t^4\,\%1\cos(t)\sin(t)$$
$$+ 2\,t^3\,\%1\cos(t)^4 - 2\,t^4\,\%1\cos(t)^3\sin(t) - t\cos(t)\sin(t\sin(t))\sin(t)$$
$$- \cos(t)^2\sin(t\sin(t))\,t^2 - 2\,t\sin(t^2\cos(t)^2) + 2\,t\sin(t^2\cos(t)^2)\cos(t)^2$$
$$- 2\sin(t)\sin(t^2\cos(t)^2)\,t^2\cos(t)dt$$
$$\%1 := \cos(t^2\cos(t)^2)$$

$$LHS = 4\,\pi$$

On the other hand, we compute the right hand side by evaluating the function at the initial and final points:

```
> A:=r&@0; B:=r&@(4*Pi);
```

$$A := [0,\, 0]$$

$$B := [4\,\pi,\, 0]$$

So the right hand side is

```
> RHS=f&@B-f&@A;
```

$$RHS = 4\,\pi$$

Notice that *Maple* took a long time to compute the left side of the F.T.C.C. and it would take you an even longer time, but it was trivial to compute the right side of the F.T.C.C.

8.2.2 Applications

Path Independence for Line Integrals An integral $\displaystyle\int_{\substack{A\\ \vec{r}(t)}}^{B} \vec{F}\cdot d\vec{s}$ is path independent in a region R if the value of the integral is the same for any curve $\vec{r}(t)$ which starts at A, ends at B and stays in the region R. If a vector field $\vec{F}$ has a scalar potential in the region R, i.e. $\vec{F} = \vec{\nabla} f$ for some function f defined in the region R, then the Fundamental Theorem of Calculus for Curves shows that $\int \vec{F}\cdot d\vec{s} = \int \vec{\nabla} f\cdot d\vec{s}$ may be computed as $f(B) - f(A)$ and so is path independent in R. Conversely, if $\int \vec{F}\cdot d\vec{s}$ is path independent in R, then the formula

$$f(P) = \int_A^P \vec{F}\cdot d\vec{s}$$

defines a scalar potential $f(P)$ in the region R. Here A is a fixed point in the region R and P is a variable point in R. If $\vec{F}$ does not have a scalar potential, then $\int \vec{F}\cdot d\vec{s}$ is not path independent and must be computed explicitly.

Example: 8.2. Compute the integral $\displaystyle\int_{(1,0,0)}^{(1,0,2\pi)} \vec{F}\cdot d\vec{s}$ for each of the following vector fields along the helix $\vec{r}(t) = (\cos t, \sin t, t)$. If the integral is path independent, you may find a potential and use the Fundamental Theorem of Calculus for Curves.

a) $\vec{F} = (2xz^2 + yz, xz, 2x^2z + xy)$

b) $\vec{G} = (yz, -xz, z^2)$

SOLUTION: a) We first enter the curve and the vector field:

```
> r:=MF(t,<cos(t)| sin(t)| t>):
> F:=MF([x,y,z],<2*x*z^2+y*z|x*z|2*x^2*z+x*y>):
```

To check if $\vec{F}$ has a scalar potential, we compute the curl and use the `SPot` command:

```
> Curl(F);
```

$$[(x,\, y,\, z) \to 0,\, (x,\, y,\, z) \to 0,\, (x,\, y,\, z) \to 0]$$

```
> f:=SPot(F);
```

$$f := (x,\, y,\, z) \to x^2\, z^2 + y\, z\, x$$

Both methods show that $\vec{F}$ has a potential, and the `SPot` command stores the potential in the variable `f`. So we evaluate the integral by using the F.T.C.C. The initial and final points are

```
> A:=<1|0|0>; B:=<1|0|2*Pi>;
```

$$A := [1, 0, 0]$$
$$B := [1, 0, 2\pi]$$

So the integral is

```
> f&@B-f&@A;
```

$$4\pi^2$$

b) The curve is the same as in part (a). The vector field is

```
> G:=MF([x,y,z],<y*z| -x*z| z^2>):
```

but it does not have a potential since

```
> g:=SPot(G);
```

$$g := \mathit{undefined}$$

So the F.T.C.C. does not apply and the integral must be done by hand. The velocity is

```
> v:=D(r) &@ t;
```

$$v := [-\sin(t), \cos(t), 1]$$

and the vector field on the curve is

```
> Gr:=G&@(r&@t);
```

$$Gr := [t\sin(t), -t\cos(t), t^2]$$

So the integral is

```
> Int(Gr &. v,t=0..2*Pi); value(%);
```

$$\int_0^{2\pi} t\,(t-1)\,dt$$
$$\frac{8}{3}\pi^3 - 2\pi^2$$

Work, Conservative Forces and Potential Energy If a particle moves along a curve $\vec{r}(t)$ under the action of a force $\vec{F}$ from a point $A = \vec{r}(a)$ to a point $B = \vec{r}(b)$, then the work done by the force on the particle is defined to be

$$Work = \int_{\substack{A \\ \vec{r}(t)}}^{B} \vec{F} \cdot d\vec{s}\ .$$

The force $\vec{F}$ is called conservative if the work integral $W$is path independent or equivalently if the force has a scalar potential, i.e. $\vec{F} = \vec{\nabla} V$. In that case the scalar potential V is called the potential energy and, by the Fundamental Theorem of Calculus for Curves, the work is the change in the potential energy:

$$Work = \int_{\substack{A \\ \vec{r}(t)}}^{B} \vec{\nabla} V \cdot d\vec{s} = V(B) - V(A)\ .$$

(See Table C.3 in Appendix C.) Note, the potential energy is only defined up to an additive constant which is sometimes fixed by requiring the potential energy to be zero at infinity.

Example: 8.3. The gravitational force of a mass M on a mass m is $\vec{F} = -\dfrac{GMm}{|\vec{r}|^3}\vec{r}$ where $\vec{r}$ is the vector from M to m. If M is fixed at the origin and m is at (x, y, z), then $\vec{r} = (x, y, z)$ and $\vec{F} = -\dfrac{GMm}{\sqrt{x^2+y^2+z^2}^3}(x, y, z)$. Find the work done in moving the mass m from $A = (a, 0, 0)$ to $B = (b, 0, 0)$ along the x-axis. Then find the work done in moving the mass m from $|\vec{r}| = \infty$ to $|\vec{r}| = R$ along an arbitrary path.

SOLUTION: We enter the force and check to see if it has a scalar potential:

```
> F:=MF([x,y,z], [-G*M*m*x/(x^2+y^2+z^2)^(3/2),
> -G*M*m*y/(x^2+y^2+z^2)^(3/2), -G*M*m*z/(x^2+y^2+z^2)^(3/2)]):
> V:=SPot(F);
```

$$V := (x, y, z) \to \frac{G\,M\,m}{\sqrt{x^2+y^2+z^2}}$$

Yes it does. So the work is path independent, the force is conservative and the potential energy is $V = \dfrac{GMm}{r}$. (Notice that the potential energy is normalized to zero at infinity.) To compute the work, we enter the initial and final points:

```
> A:=[a,0,0]: B:=[b,0,0]:
```

and find the change in potential energy:

```
> W:=V&@B-V&@A;
```

$$\mathrm{W} := \frac{G\,M\,m}{\sqrt{b^2}} - \frac{G\,M\,m}{\sqrt{a^2}}$$

You can get the square root and square to cancel by first executing `assume(a>0,b>0);` Since the potential energy only depends on the distance from the origin $|\vec{r}|$ and the work is path independent, the work in bringing a mass from $|\vec{r}| = \infty$ to $|\vec{r}| = R$ along an arbitrary path is

```
> assume(R>0); Winf:=limit(subs(b=R,W),a=infinity);
```

$$Winf := \frac{G\,M\,m}{R\tilde{}}$$

8.3 Green's Theorem

Green's Theorem. If R is a "nice" region in $\mathbb{R}^2$ and ∂R is its boundary curve traversed so that R stays on the left, and if P and Q are differentiable functions defined in a neighborhood of R, then

$$\iint_R \left(\frac{\partial Q}{\partial x} - \frac{\partial P}{\partial y}\right) dx\,dy = \oint_{\partial R} P\,dx + Q\,dy$$

We will not clarify the definition of a "nice" region. There are two variants of Green's Theorem. First, if we define the vector field $\vec{F} = (P, Q)$, then Green's Theorem may be rewritten as:

2-Dimensional Stokes' Theorem. If R is a "nice" region in $\mathbb{R}^2$ and ∂R is its boundary curve traversed so that R stays on the left, and if $\vec{F}$ is a differentiable vector field defined in a neighborhood of R, then

$$\iint_R (\vec{\nabla} \times \vec{F}) \cdot \hat{k}\,dx\,dy = \oint_{\partial R} \vec{F} \cdot d\vec{s}$$

where $\vec{\nabla} \times \vec{F}$ is computed by extending $\vec{F}$ to the 3-dimensional vector field $\vec{F} = (F_1, F_2, 0)$.

Second, if we define the vector field $\vec{G} = (Q, -P)$, then Green's Theorem may be rewritten as:

2-Dimensional Gauss' Theorem. If R is a "nice" region in $\mathbb{R}^2$ and ∂R is its boundary curve traversed so that R stays on the left, and if $\vec{G}$ is a differentiable vector field defined in a neighborhood of R, then

$$\iint_R \vec{\nabla} \cdot \vec{G}\,dx\,dy = \oint_{\partial R} \vec{G} \cdot d\vec{n}$$

where the normal vector differential is $d\vec{n} = (dy, -dx) = \left(\frac{dy}{dt}, -\frac{dx}{dt}\right) dt$.

8.3.1 Verification

Example: 8.4. Verify Green's Theorem and its variants for the indicated functions or vector field on the region between the parabola $y = x^2$ and the line $y = 4$. Notice all three cases produce the same integrals.

a) Green's Theorem with $P = -y^3$ and $Q = x^3$.

b) The 2-Dimensional Stokes' Theorem with $\vec{F} = (-y^3, x^3)$.

c) The 2-Dimensional Gauss' Theorem with $\vec{G} = (x^3, y^3)$.

NOTE: This is not a proof of the theorem because you are not verifying it for general functions and regions.

SOLUTION:

a) For the left side of Green's Theorem, we first enter P and Q:

```
> P:=-y^3:  Q:=x^3:
```

We then compute the integrand $\dfrac{\partial Q}{\partial x} - \dfrac{\partial P}{\partial y}$ and integrate:

```
> integrand:=diff(Q,x)-diff(P,y);
```

$$integrand := 3\,x^2 + 3\,y^2$$

```
> Muint(integrand,y=x^2..4,x=-2..2); LHS=value(%);
```

$$\int_{-2}^{2}\int_{x^2}^{4} 3\,x^2 + 3\,y^2\,dy\,dx$$

$$LHS = \frac{8576}{35}$$

The right side is the sum of two line integrals, the first along the curve $(x_1, y_1) = (t, t^2)$ for $-2 \le t \le 2$ and the second along the curve $(x_2, y_2) = (2 - t, 4)$ for $0 \le t \le 4$. Notice the total curve is traversed counterclockwise. For the first curve, the coordinates, their derivatives, the functions P and Q and the line integral are:

```
> x1:=t:  y1:=t^2:
> dx1, dy1:=diff(x1,t), diff(y1,t);
```

$$dx1,\ dy1 := 1,\ 2\,t$$

```
> P1, Q1:=subs(x=x1,y=y1,P), subs(x=x1,y=y1,Q);
```

$$P1,\ Q1 := -t^6,\ t^3$$

```
> Int(P1*dx1+Q1*dy1,t=-2..2); I1:=value(%);
```

$$\int_{-2}^{2} -t^6 + 2\,t^4\,dt$$

$$I1 := \frac{-384}{35}$$

For the second curve, the analogous quantities are

```
> x2:=2-t:  y2:=4:
> dx2, dy2:=diff(x2,t), diff(y2,t);
```

$$dx2,\ dy2 := -1,\ 0$$

```
> P2, Q2:=subs(x=x2,y=y2,P), subs(x=x2,y=y2,Q);
```

$$P2,\ Q2 := -64,\ (2 - t)^3$$

```
> Int(P2*dx2+Q2*dy2,t=0..4); I2:=value(%);
```

$$\int_{0}^{4} 64\,dt$$

$$I2 := 256$$

So the total line integral is:

```
> RHS=I1+I2;
```

$$RHS = \frac{8576}{35}$$

b) For the left side of the 2-Dimensional Stokes' Theorem, we enter the vector field as the 3-dimensional field $\vec{F} = (-y^3, x^3, 0)$:

```
> F:=MF([x,y,z],<-y^3|x^3|0>):
```

compute the $\hat{k}$-component of the curl and integrate:

```
> integrand:=(Curl(F) &@ <x,y,z>) &. <0,0,1>;
```

$$integrand := 3\,x^2 + 3\,y^2$$

```
> Muint(integrand,y=x^2..4,x=-2..2); LHS=value(%);
```

$$\int_{-2}^{2}\int_{x^2}^{4} 3\,x^2 + 3\,y^2\,dy\,dx$$

$$LHS = \frac{8576}{35}$$

For the right side, there are again two parts. For the first curve, the position, velocity, vector field on the curve and line integral are

```
> r1:=MF(t,<t|t^2|0>); v1:=D(r1)&@t; Fr1:=F&@(r1&@t);
```

$$r1 := [t \to t,\, t \to t^2,\, t \to 0]$$

$$v1 := [1,\, 2\,t,\, 0]$$

$$Fr1 := [-t^6,\, t^3,\, 0]$$

```
> Int(Fr1 &. v1,t=-2..2); I1:=value(%);
```

$$\int_{-2}^{2} -\,t^6 + 2\,t^4\,dt$$

$$I1 := \frac{-384}{35}$$

For the second curve, the analogous quantities are

```
> r2:=MF(t,<2-t|4|0>); v2:=D(r2)&@t; Fr2:=F&@(r2&@t);
```

$$r2 := [t \to 2 - t,\, t \to 4,\, t \to 0]$$

$$v2 := [-1,\, 0,\, 0]$$

$$Fr2 := [-64,\, (2-t)^3,\, 0]$$

```
> Int(Fr2 &. v2,t=0..4); I2:=value(%);
```

$$\int_{0}^{4} 64\,dt$$

$$I2 := 256$$

So the total line integral is again:

```
> RHS=I1+I2;
```

$$RHS = \frac{8576}{35}$$

c) For the left side of the 2-Dimensional Gauss' Theorem, we enter the vector field:

```
> G:=MF([x,y],<x^3|y^3>):
```

compute the divergence and integrate:

```
> integrand:=Div(G)(x,y);
```

$$integrand := 3\,x^2 + 3\,y^2$$

```
> Muint(integrand,y=x^2..4,x=-2..2); LHS=value(%);
```

$$\int_{-2}^{2}\int_{x^2}^{4} 3\,x^2 + 3\,y^2\,dy\,dx$$

$$LHS = \frac{8576}{35}$$

For the right side, there are again two parts. For the first curve, the position and velocity were entered in part (b). The normal vector, vector field on the curve and line integral are

```
> n1:=<v1[2] | -v1[1]>; Gr1:=G&@(r1&@t);
```

$$n1 := [2\,t,\ -1]$$

$$Gr1 := [t^3,\ t^6]$$

```
> Int(Gr1 &. n1,t=-2..2); I1:=value(%);
```

$$\int_{-2}^{2} 2\,t^4 - t^6\,dt$$

$$I1 := \frac{-384}{35}$$

For the second curve, the analogous quantities are

```
> n2:=<v2[2] | -v2[1]>; Gr2:=G&@(r2&@t);
```

$$n2 := [0,\ 1]$$

$$Gr2 := \left[(2-t)^3,\ 64\right]$$

```
> Int(Gr2 &. n2,t=0..4); I2:=value(%);
```

$$\int_{0}^{4} 64\,dt$$

$$I2 := 256$$

So the total line integral is again:

```
> RHS=I1+I2;
```

$$RHS = \frac{8576}{35}$$

Notice we have gotten the same answer six different ways.

8.3.2 Applications

Area as a Line Integral Several special cases of Green's Theorem allow one to compute the area enclosed in a closed curve as a line integral around the curve. In particular, if $P = ay$ and $Q = bx$

then $\dfrac{\partial Q}{\partial x} - \dfrac{\partial P}{\partial y} = b - a$. Hence,

$$\oint_{\partial R} ay\,dx + bx\,dy = \iint_{R} (b-a)\,dx\,dy = (b-a)\mathcal{A}rea(R)$$

Thus, making three different choices for a and b, we have

$$\mathcal{A}rea(R) = -\oint_{\partial R} y\,dx = \oint_{\partial R} x\,dy = \frac{1}{2}\oint_{\partial R} -y\,dx + x\,dy$$

NOTE: This formula for area explains the results in example 6.7.

Example: 8.5. Compute the area of each of the following regions.

a) The region between the parabola $y = x^2$ and the line $y = 4$.

b) The region inside one loop of the polar daisy $r = \sin(4\theta)$.

SOLUTION:

a) We will use the formula $\mathcal{A}rea(R) = \oint_{\partial R} -y\,dx$. So the relevant vector field is

```
> F:=MF([x,y],<-y|0>);
```

$$F := [(x, y) \to -y,\ (x, y) \to 0]$$

The boundary of the region between the parabola $y = x^2$ and the line $y = 4$ must be traversed counterclockwise. For the parabola, the position, velocity, vector field on the curve and line integral are

```
> r1:=MF(t,<t|t^2>); v1:=D(r1)&@t; Fr1:=F&@(r1&@t);
```

$$r1 := [t \to t,\ t \to t^2]$$

$$v1 := [1,\ 2t]$$

$$Fr1 := [-t^2,\ 0]$$

```
> Int(Fr1 &.  v1,t=-2..2); A1:=value(%);
```

$$\int_{-2}^{2} -t^2\,dt$$

$$A1 := \frac{-16}{3}$$

For the line, the analogous quantities are

```
> r2:=MF(t,<2-t|4>); v2:=D(r2)&@t; Fr2:=F&@(r2&@t);
```

$$r2 := [t \to 2-t,\ t \to 4]$$

$$v2 := [-1,\ 0]$$

$$Fr2 := [-4,\ 0]$$

```
> Int(Fr2 &.  v2,t=0..4); A2:=value(%);
```

$$\int_0^4 4\,dt$$

$$A2 := 16$$

So the total area is:

```
> Area=A1+A2;
```

$$Area = \frac{32}{3}$$

c) Before computing the area, we plot the polar daisy, $r = \sin(4\theta)$:

```
> polarplot(sin(4*theta), theta=0..2*Pi, scaling=constrained);
```

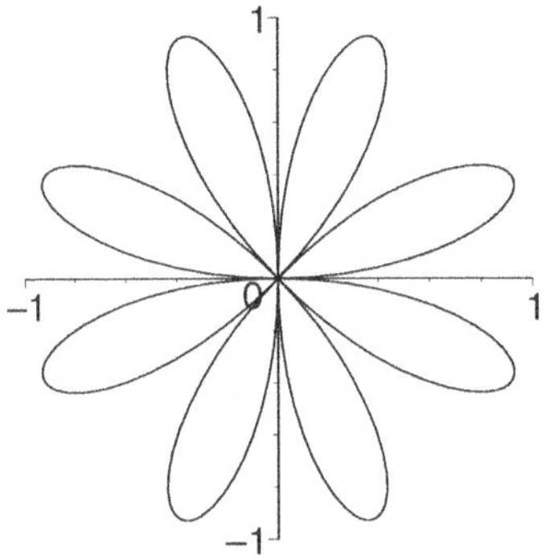

We see that one loop has the parameter range $0 \le \theta \le \frac{\pi}{4}$. Since the rectangular coordinates are related to the polar coordinates by $x = r\cos\theta, y = r\sin\theta$, the rectangular parametrization is

```
> r:=MF(theta, <sin(4*theta)*cos(theta)| sin(4*theta)*sin(theta)>);
```

$$r := [\theta \to 4\sin(\theta)\cos(\theta)^2(2\cos(\theta)^2 - 1),\ \theta \to \sin(4\,\theta)\sin(\theta)]$$

and the velocity is

```
> v:=D(r)&@theta;
```

$$v := \Big[4\cos(\theta)^3(2\cos(\theta)^2-1) - 8\sin(\theta)^2\cos(\theta)(2\cos(\theta)^2-1) - 16\sin(\theta)^2\cos(\theta)^3,$$
$$4\cos(4\,\theta)\sin(\theta) + \sin(4\,\theta)\cos(\theta)\Big]$$

We will use the formula $\mathcal{A}rea(R) = \oint_{\partial R} x\,dy$. So the relevant vector field is

```
> F:=MF([x,y], <0| x>):
```

and its value on the curve is

```
> Fr:=F&@(r&@theta);
```

$$Fr := [0,\ 4\sin(\theta)\cos(\theta)^2(2\cos(\theta)^2-1)]$$

So the area is

```
> Int(Fr &. v,theta=0..Pi/4); Area=value(%);
```

$$\int_0^{\frac{\pi}{4}} 16\sin(\theta)^2\cos(\theta)^2\,(11\cos(\theta)^2 + 20\cos(\theta)^6 - 28\cos(\theta)^4 - 1)\,d\theta$$

$$Area = \frac{\pi}{16}$$

Obviously, you would not like to do this integral by hand.

8.4 Stokes' Theorem (The Curl Theorem)

Stokes' Theorem. If S is a "nice" parametrized surface in $\mathbb{R}^3$ and ∂S is its boundary curve traversed so that the normal to the surface and the velocity of the curve are related by the right hand rule, and if $\vec{F}$ is a differentiable vector field defined in a neighborhood of S, then

$$\iint_S (\vec{\nabla} \times \vec{F}) \cdot d\vec{S} = \oint_{\partial S} \vec{F} \cdot d\vec{s}$$

We will not clarify the definition of a "nice" region.

8.4.1 Verification

Example: 8.6. Verify Stokes' Theorem by computing both sides for the vector field $\vec{F} = (yz, -xz, xyz)$ and the surface S which is the piece of the hyperbolic paraboloid $z = x^2 - y^2$ above the rectangle $-2 \le x \le 2$ and $-3 \le y \le 3$ with normal pointing toward increasing z.

NOTE: This is not a proof of the theorem because you are not verifying it for general vector fields and surfaces.

SOLUTION: The hyperbolic paraboloid $z = x^2 - y^2$ is above a square. So we use rectangular coordinates to parametrize the surface:

```
> R:=MF([u,v], <u| v| u^2 - v^2>):
```

NOTE: We could have used the letters x and y as the parameters instead of u and v, but that is sometimes confusing.

The tangent and normal vectors are:

```
> Ru:=D[1](R) &@ <u,v>; Rv:=D[2](R) &@ <u,v>; N:=Ru &x Rv;
```

$$Ru := [1,\ 0,\ 2\,u]$$

$$Rv := [0,\ 1,\ -2\,v]$$

$$N := [-2\,u,\ 2\,v,\ 1]$$

Since the z-component of $\vec{N}$ is positive, it points up as required. Next, we enter the vector field and compute its curl:

```
> F:=MF([x,y,z],<y*z| -x*z| x*y*z>):
> curlF:=Curl(F);
```

$$curlF := [(x, y, z) \to x\,z + x,\ (x, y, z) \to y - y\,z,\ (x, y, z) \to -2\,z]$$

Then we evaluate the curl on the surface, dot into the normal and integrate to obtain the left hand side of Stokes' Theorem:

```
> curlFR:=curlF&@(R &@ <u,v>);
```

$$curlFR := [u\,(u^2 - v^2) + u,\ v - v\,(u^2 - v^2),\ -2\,u^2 + 2\,v^2]$$

```
> Muint(curlFR &. N, u=-2..2, v=-3..3); LHS=value(%);
```

$$\int_{-3}^{3}\int_{-2}^{2} - 2\,u^4 - 4\,u^2 + 4\,v^2 + 2\,v^4\,du\,dv$$

$$LHS = 784$$

The boundary of the hyperbolic paraboloid consists of four curves. We parametrize them so they are traversed counterclockwise as seen from the positive z-axis and compute the velocities: (We suppress the output.)

```
> r1:=MF(u,<u| -3| u^2-9>):  v1:=D(r1) &@ u:  #for u=-2..2
> r2:=MF(v,<2| v| 4-v^2>):  v2:=D(r2) &@ v:  #for v=-3..3
> r3:=MF(t,<-t| 3| t^2-9>):  v3:=D(r3) &@ t:  #for t=-2..2
> r4:=MF(t,<-2| -t| 4-t^2>):  v4:=D(r4) &@ t:  #for t=-3..3
```

Then we restrict $\vec{F}$ to each curve, dot into the velocity and integrate: (To save space, we display the answers on one line.)

```
> I1:=int(F&@(r1 &@ u) &.  v1, u=-2..2):
> I2:=int(F&@(r2 &@ v) &.  v2, v=-3..3):
> I3:=int(F&@(r3 &@ t) &.  v3, t=-2..2):
> I4:=int(F&@(r4 &@ t) &.  v4, t=-3..3):
> I1, I2, I3, I4;
```

$$\frac{1516}{5},\ \frac{444}{5},\ \frac{1516}{5},\ \frac{444}{5}$$

The sum of these integrals is the right side of Stokes'Theorem:

```
> RHS=I1 + I2 + I3 + I4;
```

$$RHS = 784$$

NOTE: The computation of these integrals can be automated as follows using the VecCalc command liv which is the alias for LineIntegralVector. This command automatically computes the velocity, evaluates the vector field on the curve, takes the dot product and computes the integral to obtain the flow along that edge. (See the help pages for Lis, lis, Liv and liv which display and compute line integrals of scalars and vectors.)

```
> I1,I2,I3,I4:=liv(F,r1, u=-2..2), liv(F,r2, v=-3..3), liv(F,r3,
> t=-2..2), liv(F,r4, t=-3..3);
```

$$I1,\ I2,\ I3,\ I4 := \frac{1516}{5},\ \frac{444}{5},\ \frac{1516}{5},\ \frac{444}{5}$$

8.4.2 Applications

Surface Integrals as Line Integrals and Line Integrals as Surface Integrals If an exercise asks you to use Stokes' Theorem to do a surface integral, it really means you are to do a line integral. On the other hand, if an exercise asks you to use Stokes' Theorem to do a line integral, it really means you are to do a surface integral.

Example: 8.7. Use Stokes' Theorem to compute the surface integral $\iint_P \vec{G} \cdot d\vec{S}$ where $\vec{G} = (-xz, -yz, x^2 + y^2 + z^2)$ and P is the paraboloid $z = x^2 + y^2$ for $z \le 9$ with normal pointing in and up.

SOLUTION: The statement of the problem really means we are to find a vector potential and do a line integral of the vector potential around the boundary curve. So we first enter the vector field $\vec{G}$ and find a vector potential:

```
> G:=MF([x,y,z], <-x*z| -y*z| x^2+y^2+z^2>):
> A:=VPot(G);
```

$$A := \left[(x, y, z) \to -\frac{1}{2} y z^2 - x^2 y - \frac{1}{3} y^3, (x, y, z) \to \frac{1}{2} x z^2, (x, y, z) \to 0\right]$$

Then by Stokes' Theorem, we have

$$\iint_P \vec{G} \cdot d\vec{S} = \iint_P (\vec{\nabla} \times \vec{A}) \cdot d\vec{S} = \oint_{\partial P} \vec{A} \cdot d\vec{s}$$

The boundary of the paraboloid is the circle $x^2+y^2 = 9$ in the plane $z = 9$ traversed counterclockwise as seen from the positive z-axis. So we enter the curve and compute the velocity:

```
> r:=MF(t,<3*cos(t)| 3*sin(t)| 9>):
> v:=D(r)&@t;
```

$$v := [-3\sin(t), 3\cos(t), 0]$$

Finally, we evaluate the vector potential on the curve and integrate:

```
> Ar:=A&@(r&@t);
```

$$Ar := \left[-\frac{243}{2}\sin(t) - 27\cos(t)^2\sin(t) - 9\sin(t)^3, \frac{243}{2}\cos(t), 0\right]$$

```
> Int(Ar &. v, t=0..2*Pi); value(%);
```

$$\int_0^{2\pi} 27\cos(t)^2 + \frac{783}{2} - 54\cos(t)^4\, dt$$

$$\frac{1539\pi}{2}$$

Example: 8.8. Use Stokes' Theorem to compute the line integral $\oint_{\partial T} \vec{F} \cdot d\vec{s}$ of the vector field $\vec{F} = (xy^2 + zy, 2xyz, yz^2 - xy)$ along the line segments from $(2, 0, 0)$ to $(0, 3, 0)$ to $(0, 0, 4)$ and back to $(2, 0, 0)$.

SOLUTION: Notice that the three line segments form the boundary of the triangle T with vertices $(2,0,0)$, $(0,3,0)$ and $(0,0,4)$ and normal pointing up into the first octant. So the statement of the problem really means we are to compute the surface integral $\iint_T (\vec{\nabla} \times \vec{F}) \cdot d\vec{S}$.

We first need to parametrize the plane of the triangle. The vertices are:

```
> A:=<2|0|0>:  B:=<0|3|0>:  C:=<0|0|4>:
```

The tangent vectors are:

```
> AB:=B-A; AC:=C-A;
```

$$AB := [-2,\ 3,\ 0]$$
$$AC := [-2,\ 0,\ 4]$$

Then the parametrized plane is $X = A + s\overrightarrow{AB} + t\overrightarrow{AC}$ and the triangle corresponds to $0 \le s \le 1$ and $0 \le t \le 1 - s$. We enter this as

```
> R:=MF([s,t], A + s*AB + t*AC);
```

$$R := [(s,\ t) \to 2 - 2\,s - 2\,t,\ (s,\ t) \to 3\,s,\ (s,\ t) \to 4\,t]$$

Notice that the tangent vectors are just $\overrightarrow{AB}$ and $\overrightarrow{AC}$. So the normal vector is:

```
> N:=AB &x AC;
```

$$N := [12,\ 8,\ 6]$$

which points into the first octant as required. Now the vector field and its curl are:

```
> F:=MF([x,y,z], <x*y^2+z*y| 2*x*y*z| y*z^2-x*y>):
> curlF:=Curl(F);
```

$$curlF := [(x,\ y,\ z) \to z^2 - x - 2\,y\,x,\ (x,\ y,\ z) \to 2\,y,\ (x,\ y,\ z) \to 2\,y\,z - 2\,y\,x - z]$$

and the restriction of the curl to the plane of the triangle is:

```
> curlFR:=curlF&@(R&@ <s,t>);
```

$$curlFR := [16\,t^2 - 2 + 2\,s + 2\,t - 6\,s\,(2 - 2\,s - 2\,t),\ 6\,s,\ 24\,s\,t - 6\,s\,(2 - 2\,s - 2\,t) - 4\,t]$$

Finally its integral over the triangle is:

```
> Muint(curlFR &. N, t=0..1-s, s=0..1); value(%);
```

$$\int_0^1 \int_0^{1-s} 192\,t^2 - 24 - 144\,s + 216\,s^2 + 360\,s\,t\,dt\,ds$$

$$13$$

Surface Independence for Surfaces Integrals A surface integral $\iint_S \vec{F} \cdot d\vec{S}$ is surface independent in a region R if the value of the integral is the same for any surface S which stays in the region R and has the same boundary curve ∂S. If the vector field $\vec{F}$ has a vector potential in the region R, i.e. $\vec{F} = \vec{\nabla} \times \vec{A}$ for some vector field $\vec{A}$ defined in the region R, then Stokes' Theorem shows that $\iint_S \vec{F} \cdot d\vec{S} = \iint_S \vec{\nabla} \times \vec{A} \cdot d\vec{S}$ may be computed as the line integral $\oint_{\partial S} \vec{A} \cdot d\vec{s}$ and so is surface independent in R. If $\vec{F}$ does not have a vector potential, then $\iint_S \vec{F} \cdot d\vec{S}$ must be computed explicitly.

Example: 8.9. Consider the surface S which is the graph $z = \sin(x)\sin(y)$ for $0 \le x \le \pi$ and $0 \le y \le \pi$ with normal pointing up. Compute $\iint_S \vec{F} \cdot d\vec{S}$ over the surface S for each of the following vector fields. If the integral is surface independent, you may use Stokes' Theorem and change the surface.

a) $\vec{F} = (y - xz, x - yz, x^2 + y^2 + z^2)$

b) $\vec{F} = (xz - y, yz - x, x^2 + y^2 + z^2)$

SOLUTION:

a) We first enter the vector field:

```
> F:=MF([x,y,z], <y-x*z| x-y*z| x^2 + y^2 + z^2>):
```

To check if $\vec{F}$ has a vector potential, we compute the divergence:

```
> Div(F);
```

$$(x, y, z) \to 0$$

So $\vec{F}$ has a vector potential, and Stokes' Theorem says

$$\iint_S \vec{F} \cdot d\vec{S} = \iint_S \vec{\nabla} \times \vec{A} \cdot d\vec{S} = \oint_{\partial S} \vec{A} \cdot d\vec{s}$$

However, the boundary of S, is also the boundary of the square $0 \le x \le \pi$ and $0 \le y \le \pi$ in the xy-plane. Let T denote this square with normal pointing up. Then Stokes' Theorem also says

$$\oint_{\partial S} \vec{A} \cdot d\vec{s} = \oint_{\partial T} \vec{A} \cdot d\vec{s} = \iint_T \vec{\nabla} \times \vec{A} \cdot d\vec{S} = \iint_T \vec{F} \cdot d\vec{S}$$

In other words, $\iint_S \vec{F} \cdot d\vec{S}$ is surface independent and it is easier to compute $\iint_T \vec{F} \cdot d\vec{S}$. The square may be parametrized as

```
> R:=MF([x,y],[x,y,0]):
```

Its tangent vectors are $\vec{R}_x = (1, 0, 0)$ and $\vec{R}_y = (0, 1, 0)$ and its normal vector is $\vec{N} = (0, 0, 1)$. With these parameters, $\vec{F}$ may be evaluated on the square by setting $z = 0$ and its dot product with $\vec{N}$ is just F_3. So the desired integral is:

```
> Muint(F[3](x,y,0), x=0..Pi, y=0..Pi); value(%);
```

$$\int_0^\pi \int_0^\pi x^2 + y^2 \, dx \, dy$$

$$\frac{2\,\pi^4}{3}$$

NOTE: Notice we never used the vector potential $\vec{A}$. So there was no reason to use the VPot command.

b) Once again we enter the vector field:

```
> F:=MF([x,y,z], <x*z-y| y*z-x| x^2 + y^2 + z^2>):
```

and check if $\vec{F}$ has a vector potential:

```
> Div(F);
```

$$(x,\, y,\, z) \to 4\, z$$

It does not! So we cannot use Stokes' Theorem. The surface S may be parametrized as

```
> R:=MF([x,y],<x|y|sin(x)*sin(y)>):
```

and so its tangent and normal vectors are

```
> Rx:=D[1](R)&@ <x,y>; Ry:=D[2](R)&@ <x,y>; N:=Rx &x Ry;
```

$$Rx := [1,\, 0,\, \cos(x)\sin(y)]$$

$$Ry := [0,\, 1,\, \sin(x)\cos(y)]$$

$$N := [-\cos(x)\sin(y),\, -\sin(x)\cos(y),\, 1]$$

On the surface, the vector field becomes

```
> FR:=F&@(R&@ <x,y>);
```

$$FR := [x\sin(x)\sin(y) - y,\, y\sin(x)\sin(y) - x,\, x^2 + y^2 + \sin(x)^2\sin(y)^2]$$

So the integral is

```
> Muint(FR &.  N, x=0..Pi, y=0..Pi); value(%);
```

$$\int_0^\pi \int_0^\pi - \cos(x)\, x\sin(x) + \cos(x)\, x\sin(x)\cos(y)^2 + \cos(x)\sin(y)\, y - \cos(y)\, y\sin(y) + \cos(y)\, y\sin(y)\cos(x)^2 + \sin(x)\cos(y)\, x + x^2 + y^2 + 1 - \cos(y)^2 - \cos(x)^2 + \cos(x)^2\cos(y)^2 dxdy$$

$$\frac{1}{2}\pi^2 + \frac{2}{3}\pi^4$$

Circulation and Flux In subsection 6.1.4, we discussed the circulation of a vector field and in subsection 6.2.6, we discussed the flux of a vector field. (See Table C.3 in Appendix C.) These are not directly related by Stokes' Theorem. Rather, the circulation of a vector field $\vec{v}$ is equal to the flux of its curl:

$$\mathit{Circulation} = \oint_{\partial S} \vec{v} \cdot d\vec{s} = \iint_S (\vec{\nabla} \times \vec{v}) \cdot d\vec{S}\ .$$

On the other hand, if the vector field $\vec{v}$ has a vector potential $\vec{A}$, then the flux of $\vec{v}$ is equal to the circulation of its vector potential:

$$\mathcal{F}lux = \iint_S \vec{v} \cdot d\vec{S} = \iint_S (\vec{\nabla} \times \vec{A}) \cdot d\vec{S} = \oint_{\partial S} \vec{A} \cdot d\vec{s}$$

Example: 8.10. Find the circulation of the vector field $\vec{F} = (-y^3 z, x^3 z, z^4)$ counterclockwise around the circle $x^2 + y^2 = 4$ in the plane $z = 5$.

SOLUTION: We could explicitly compute the circulation as the line integral $\oint_C \vec{F} \cdot d\vec{s}$. However we will use Stokes' Theorem and compute the circulation as the surface integral $\iint_S (\vec{\nabla} \times \vec{F}) \cdot d\vec{S}$. So we enter the vector field and compute the curl:

```
> F:=MF([x,y,z], <-y^3*z| x^3*z| z^4>):
> curlF:=Curl(F);
```

$$curlF := [(x, y, z) \to -x^3, (x, y, z) \to -y^3, (x, y, z) \to 3\,x^2\,z + 3\,y^2\,z]$$

We parametrize the disk and compute the tangent and normal vectors:

```
> R:=MF([r, theta], <r*cos(theta)| r*sin(theta)| 5>):
> Rr:=D[1](R) &@ <r,theta>:  Rtheta:=D[2](R) &@ <r,theta>:
> N:=Rr &x Rtheta;
```

$$N := [0, 0, r]$$

Since the circle is traversed counterclockwise, the normal to the disk should point up as it does. Finally we evaluate $\vec{\nabla} \times \vec{F}$ on the surface, dot into the normal and integrate:

```
> Muint(curlF&@(R&@ <r,theta>) &.  N, r=0..2, theta=0..2*Pi);
> Circ:=value(%);
```

$$\int_0^{2\pi} \int_0^2 15\,r^3\,dr\,d\theta$$

$$Circ := 120\,\pi$$

Example: 8.11. Find the flux of the vector field $\vec{F} = \left(xy^2, yx^2, -z(x^2 + y^2)\right)$ upward through the paraboloid $z = x^2 + y^2$ for $z \le 9$.

SOLUTION: We could explicitly compute the flux as the surface integral $\iint_P \vec{F} \cdot d\vec{S}$. However we will use Stokes' Theorem and compute the flux as the line integral $\oint_{\partial P} \vec{A} \cdot d\vec{s}$ where $\vec{A}$ is a vector potential for $\vec{F}$. So we enter the vector field and find its vector potential:

```
> F:=MF([x,y,z],<x*y^2| y*x^2| -z*(x^2+y^2)>):
> A:=VPot(F);
```

$$A := [(x, y, z) \to x^2\, y\, z,\ (x, y, z) \to -x\, y^2\, z,\ (x, y, z) \to 0]$$

We now parametrize the boundary circle. Since the paraboloid is oriented upward, the circle should be traversed counterclockwise:

```
> r:=MF(theta, <3*cos(theta)| 3*sin(theta)| 9>):
```

Next we compute the velocity and restrict the vector potential to the circle:

```
> v:=D(r) &@ theta; Ar:=A&@(r&@ theta);
```

$$v := [-3\sin(\theta),\ 3\cos(\theta),\ 0]$$

$$Ar := [243\cos(\theta)^2\sin(\theta),\ -243\cos(\theta)\sin(\theta)^2,\ 0]$$

So the flux is:

```
> Int(Ar &.  v, theta=0..2*Pi); Flux=value(%);
```

$$\int_0^{2\pi} -1458\cos(\theta)^2\sin(\theta)^2\, d\theta$$

$$Flux = -\frac{729\,\pi}{2}$$

8.5 Gauss' Theorem (The Divergence Theorem)

Gauss' Theorem. If V is a "nice" solid region in $\mathbb{R}^3$ and ∂V is its boundary surface oriented with the normal pointing out from the volume, and if $\vec{F}$ is a differentiable vector field defined in a neighborhood of V, then

$$\iiint\limits_V \vec{\nabla}\cdot\vec{F}\, dV = \iint\limits_{\partial V} \vec{F}\cdot d\vec{S}$$

We will not clarify the definition of a "nice" region.

8.5.1 Verification

Example: 8.12. Verify Gauss' Theorem by computing both sides for the vector field $F = (x^3z^2, y^3z^2, z^3)$ and the solid region V above the paraboloid $z = x^2 + y^2$ and below the plane $z = 9$.

NOTE: This is not a proof of the theorem because you are not verifying it for general vector fields and volumes.

SOLUTION: We enter the vector field and compute the divergence:

```
> F:=MF([x,y,z], <x^3*z^2| y^3*z^2| z^3>):
> divF:=Div(F);
```

$$divF := (x, y, z) \to 3\,x^2\,z^2 + 3\,y^2\,z^2 + 3\,z^2$$

It is easiest to integrate over the solid paraboloid in cylindrical coordinates. So we define cylindrical coordinates and evaluate the divergence in these coordinates.

```
> R:=MF([r,theta,z], <r*cos(theta)| r*sin(theta)| z>):
> divFcyl:=simplify(divF&@(R &@ <r,theta,z>));
```

$$divFcyl := 3\,z^2\,(1+r^2)$$

Then we compute the integral. (Don't forget the Jacobian!)

```
> Muint(divFcyl*r, z=r^2..9, r=0..3, theta=0..2*Pi); LHS=value(%);
```

$$\int_0^{2\pi}\int_0^3\int_{r^2}^9 3\,z^2\,(1+r^2)\,r\,dz\,dr\,d\theta$$

$$LHS = \frac{452709\,\pi}{20}$$

There are two boundary surfaces. First, the paraboloid may be parametrized as

```
> R1:=MF([r,theta], <r*cos(theta)| r*sin(theta)| r^2>):
```

The tangent and normal vectors are:

```
> R1r:=D[1](R1)&@ <r,theta>; R1theta:=D[2](R1)&@ <r,theta>;
> N1:=R1r &x R1theta;
```

$$R1r := [\cos(\theta),\ \sin(\theta),\ 2\,r]$$

$$R1theta := [-r\sin(\theta),\ r\cos(\theta),\ 0]$$

$$N1 := [-2\,r^2\cos(\theta),\ -2\,r^2\sin(\theta),\ r]$$

This normal points up, but the outward normal should point down. So we reverse the normal:

```
> N1:=-N1;
```

$$N1 := [2\,r^2\cos(\theta),\ 2\,r^2\sin(\theta),\ -r]$$

The restriction of $\vec{F}$ to the paraboloid is

```
> FR1:=F&@(R1&@ <r,theta>);
```

$$FR1 := [r^7\cos(\theta)^3,\ r^7\sin(\theta)^3,\ r^6]$$

So the integral is

```
> Muint(FR1 &. N1, r=0..3, theta=0..2*Pi); I1:=value(%);
```

$$\int_0^{2\pi}\int_0^3 r^7\,(4\,r^2\cos(\theta)^4 - 1 + 2\,r^2 - 4\,r^2\cos(\theta)^2)\,dr\,d\theta$$

$$I1 := \frac{321489\,\pi}{20}$$

Second, the plane may be parametrized by

```
> R2:=MF([r,theta], <r*cos(theta)| r*sin(theta)| 9>):
```

The tangent and normal vectors are:

```
> R2r:=D[1](R2)&@ <r,theta>; R2theta:=D[2](R2)&@ <r,theta>;
> N2:=R2r &x R2theta;
```

$$R2r := [\cos(\theta),\ \sin(\theta),\ 0]$$

$$R2theta := [-r\sin(\theta),\ r\cos(\theta),\ 0]$$

$$N2 := [0,\ 0,\ r]$$

This time the normal points up as it should. The restriction of $\vec{F}$ to the plane is

```
> FR2:=F&@(R2&@ <r,theta>);
```

$$FR2 := [81\, r^3 \cos(\theta)^3,\ 81\, r^3 \sin(\theta)^3,\ 729]$$

So the integral is

```
> Muint(FR2 &. N2, r=0..3, theta=0..2*Pi); I2:=value(%);
```

$$\int_0^{2\pi} \int_0^3 729\, r\, dr\, d\theta$$

$$I2 := 6561\,\pi$$

So the total integral over the boundary is:

```
> RHS=I1+I2;
```

$$RHS = \frac{452709\,\pi}{20}$$

8.5.2 Applications

Surface Integrals as Volume Integrals If an exercise asks you to use Gauss' Theorem to do a surface integral over a closed surface, it really means you are to do a volume integral.

Example: 8.13. Use Gauss' Theorem to compute the following surface integrals over the complete surface of the cylinder C given by $x^2 + y^2 \le 4$ for $0 \le z \le 5$ with normal pointing out.

a) $\displaystyle\iint_{\partial C} \vec{F} \cdot d\vec{S}$ for $F = (x^3 z, y^3 z, x^2 + y^2 + z^2)$.

b) $\displaystyle\iint_{\partial C} \vec{G} \cdot d\vec{S}$ for $G = (-xz, -yz, x^2 + y^2 + z^2)$.

SOLUTION:

a) We enter the vector field and compute the divergence:

```
> F:=MF([x,y,z], <x^3*z| y^3*z| x^2+y^2+z^2>):
> divF:=Div(F);
```

$$divF := (x,\ y,\ z) \to 3\,x^2\, z + 3\,y^2\, z + 2\,z$$

Then we evaluate the divergence in cylindrical coordinates and integrate:

```
> R:=MF([r,theta,z], <r*cos(theta)| r*sin(theta)| z>):
> divFcyl:=simplify(divF&@(R &@ <r,theta,z>));
```

$$divFcyl := z\,(2 + 3\,r^2)$$

```
> Muint(divFcyl*r, r=0..2, theta=0..2*Pi, z=0..5); value(%);
```

$$\int_0^5 \int_0^{2\pi} \int_0^2 z\,(2+3\,r^2)\,r\,dr\,d\theta\,dz$$

$$400\,\pi$$

b) We enter the vector field and compute the divergence:

```
> G:=MF([x,y,z], [-x*z, -y*z, x^2+y^2+z^2]):
> divG:=Div(G);
```

$$divG := (x,\, y,\, z) \to 0$$

Since the divergence is zero, the integral is automatically zero.

Volume as a Surface Integral A special case of Gauss' Theorem allows one to compute the volume enclosed in a closed surface as a surface integral. In particular, for the general position vector $\vec{r} = (x, y, z)$ the divergence is $\vec{\nabla} \cdot \vec{r} = 3$. Hence,

$$\iint\limits_{\partial V} \vec{r} \cdot d\vec{S} = \iiint\limits_{V} 3\,dV = 3\,\mathcal{V}olume(V)$$

Thus

$$\mathcal{V}olume(V) = \frac{1}{3} \iint\limits_{\partial V} \vec{r} \cdot d\vec{S}$$

NOTE: This formula for volume explains the results in exercise 6.13.

Example: 8.14. Use a surface integral to compute the volume of the region between the paraboloid $z = x^2 + y^2$ and the plane $z = 9$.

SOLUTION: As in example 8.12, there are two boundary surfaces. For the paraboloid, the parametrization and normal vector are:

```
> R1 &@ <r,theta>; N1;
```

$$[r\cos(\theta),\, r\sin(\theta),\, r^2]$$

$$[2\,r^2\cos(\theta),\, 2\,r^2\sin(\theta),\, -r]$$

In the integral $\frac{1}{3} \iint\limits_{\partial V} \vec{r} \cdot d\vec{S}$ the vector field is $\frac{1}{3}$ of the position vector R1. So the first surface integral is

```
> 1/3*Muint((R1 &@ <r,theta>) &.  N1, r=0..3, theta=0..2*Pi);
> V1:=value(%);
```

$$\frac{1}{3} \int_0^{2\pi} \int_0^3 r^3\,dr\,d\theta$$

$$V1 := \frac{27\,\pi}{2}$$

For a disk in the plane $z = 9$, the parametrization and normal vector are:

```
> R2 &@ <r,theta>; N2;
```

$$[r\cos(\theta),\ r\sin(\theta),\ 9]$$
$$[0,\ 0,\ r]$$

So the second surface integral is

```
> 1/3*Muint((R2 &@ <r,theta>) &. N2, r=0..3, theta=0..2*Pi);
> V2:=value(%);
```

$$\frac{1}{3}\int_0^{2\pi}\int_0^3 9\,r\,dr\,d\theta$$
$$V2 := 27\,\pi$$

So the volume is:

```
> Volume=V1+V2;
```

$$Volume = \frac{81\,\pi}{2}$$

Expansion, Divergence and Source The expansion of a vector field was introduced in subsection 6.2.6. By Gauss' Theorem, the expansion of a vector field out of a surface is the integral of its divergence over the enclosed volume:

$$\mathcal{E}xpansion = \iint_{\partial V} \vec{v}\cdot d\vec{S} = \iiint_{V} \vec{\nabla}\cdot\vec{v}\,dV\ .$$

(See Table C.3 in Appendix C.) If the expansion is interpreted as the amount of stuff flowing out of the surface, then the divergence, $\vec{\nabla}\cdot\vec{v}$, should be interpreted as the amount of stuff spreading out from a point. Then its integral over a volume V is again the net amount of stuff which is flowing out of the volume.

The negative of the expansion is called the contraction; the negative of the divergence is called the convergence. If the expansion is positive, we say the stuff is expanding out of the volume; if the expansion is negative, we say the stuff is contracting. If the divergence is positive at a point then we say the point is a source for the stuff; if the divergence is negative at a point then we say the point is a sink for the stuff.

Example: 8.15. Consider the velocity field $\vec{v} = (x^3 z, y^3 z, \frac{3}{4}z^4)$.

a) Locate the sources and sinks of the fluid.

b) Find the expansion of the fluid out of the cube $-1 \le x \le 2$, $-1 \le y \le 2$, $-1 \le z \le 2$ by two methods. Is the fluid expanding or contracting out of the cube? Identify the faces of the cube on which the fluid is flowing in or out of the cube.

SOLUTION:

a) We first enter the velocity field and compute the divergence:

```
> v:=MF([x,y,z], <x^3*z| y^3*z| 3/4*z^4>):
```

```
> div_v:=Div(v) &@ <x,y,z>; div_v:=factor(%);
```

$$div\_v := 3\,x^2\,z + 3\,y^2\,z + 3\,z^3$$

$$div\_v := 3\,z\,(x^2 + y^2 + z^2)$$

So the divergence is positive when $z > 0$ and negative when $z < 0$. Thus the points above the xy-plane are sources and the points below the xy-plane are sinks.

b) We first compute the expansion using Gauss' Theorem by integrating the divergence over the cube:

```
> Muint(div_v, x=-1..2, y=-1..2, z=-1..2); Expansion=value(%);
```

$$\int_{-1}^{2}\int_{-1}^{2}\int_{-1}^{2} 3\,z\,(x^2 + y^2 + z^2)\,dx\,dy\,dz$$

$$Expansion = \frac{729}{4}$$

Since the expansion is positive, the fluid is expanding out of the cube.

Next we compute the expansion explicitly by integrating the velocity over each face. We parametrize each face (carefully choosing the order of the parameters so the normal will point outward) and use the `VecCalc` command `siv` which is the alias for `SurfaceIntegralVector`. This command automatically computes the tangent vectors and normal vector for the surface, evaluates the vector field on the surface, takes the dot product and computes the integral to obtain the flux out of that face. (See the help pages for `Sis`, `sis`, `Siv` and `siv` which display and calculate surface integrals of scalars and vectors.) To save space, we display the answers on one line.

```
> R1:=MF([z,y], <-1,y,z>):  F1:=siv(v, R1, y=-1..2, z=-1..2):
> R2:=MF([y,z], <2,y,z>):  F2:=siv(v, R2, y=-1..2, z=-1..2):
> R3:=MF([x,z], <x,-1,z>):  F3:=siv(v, R3, x=-1..2, z=-1..2):
> R4:=MF([z,x], <x,2,z>):  F4:=siv(v, R4, x=-1..2, z=-1..2):
> R5:=MF([y,x], <x,y,-1>):  F5:=siv(v, R5, x=-1..2, y=-1..2):
> R6:=MF([x,y], <x,y,2>):  F6:=siv(v, R6, x=-1..2, y=-1..2):
> F1, F2, F3, F4, F5, F6;
```

$$\frac{9}{2},\ 36,\ \frac{9}{2},\ 36,\ \frac{-27}{4},\ 108$$

Thus the flux for each face is positive except for F_5. So the fluid is flowing out of all of the faces of the cube except the bottom face where $z = -1$. Finally, we check that the total flux out of the cube is equal to the expansion :

```
> TotalFlux=F1+F2+F3+F4+F5+F6;
```

$$TotalFlux = \frac{729}{4}$$

8.6 Related Line, Surface and Volume Integrals

8.6.1 Related Line and Surface Integrals

Suppose C_1 and C_2 are two *open* curves which start at A and end at B and stay in a region R. We say that C_1 can be continuously deformed into C_2 within R if there is a surface S within R whose boundary is $\partial S = C_2 - C_1$. This means that the boundary of S may be traversed with the proper orientation by travelling forward along C_2 and then backward along C_1.

Similarly, suppose C_1 and C_2 are two *closed* curves which stay in a region R. We say that C_1 can be continuously deformed into C_2 within R if there is a surface S within R whose boundary is $\partial S = C_2 - C_1$. This means that the boundary of S consists of the two curves C_1 and C_2 with C_2 traversed forwards and C_1 traversed backwards.

In either case, $\partial S = C_2 - C_1$ and so Stokes' Theorem says

$$\iint_S \vec{\nabla} \times \vec{F} \cdot d\vec{S} = \oint_{\partial S} \vec{F} \cdot d\vec{s} = \int_{C_2} \vec{F} \cdot d\vec{s} - \int_{C_1} \vec{F} \cdot d\vec{s}$$

This may be rewritten as

$$\int_{C_2} \vec{F} \cdot d\vec{s} = \int_{C_1} \vec{F} \cdot d\vec{s} + \iint_S \vec{\nabla} \times \vec{F} \cdot d\vec{S}$$

In other words, to compute $\int_{C_2} \vec{F} \cdot d\vec{s}$, you may alternatively compute $\int_{C_1} \vec{F} \cdot d\vec{s}$ and $\iint_S \vec{\nabla} \times \vec{F} \cdot d\vec{S}$, if that is easier.

In the special case that $\vec{\nabla} \times \vec{F} = 0$ everywhere in the region R, then

$$\int_{C_2} \vec{F} \cdot d\vec{s} = \int_{C_1} \vec{F} \cdot d\vec{s}$$

whenever C_1 can be continuously deformed into C_2 within R. For the case of open curves, this is a special case of the path independence discussed at the beginning of section 8.2.2.

Example: 8.16. For each of the following vector fields, compute the line integral $\int_C \vec{F} \cdot d\vec{s}$ along the curve C which consists of the three line segments from $(0, 0, 0)$ to $(0, 0, \pi)$ to (π, π, π) to $(\pi, \pi, 0)$.

a) $\vec{F} = (\sin(x), \sin(y), \sin(z))$.

b) $\vec{F} = (\sin(y) - \sin(z), \sin(x) - \sin(z), \cos(x) + \cos(y))$.

SOLUTION:

a) We first enter the vector field and compute the curl:

```
> F:=MF([x,y,z], <sin(x)| sin(y)| sin(z)>):
> Curl(F);
```

$$[(x, y, z) \to 0, (x, y, z) \to 0, (x, y, z) \to 0]$$

Since the curl is zero, we can replace the integral along C by a integral along the single line segment L from $(0,0,0)$ to $(\pi,\pi,0)$. The parametrized line and the velocity are

```
> r:=MF(t, <t|t|0>): v:=D(r)&@t;
```

$$v := [1, 1, 0]$$

Along the curve the vector field is:

```
> Fr:=F&@(r&@t);
```

$$Fr := [\sin(t), \sin(t), 0]$$

So the integral is

```
> Int(Fr &. v, t=0..Pi); value(%);
```

$$\int_0^\pi 2\sin(t)\,dt$$

$$4$$

b) We enter the vector field and compute the curl:

```
> F:=MF([x,y,z], <sin(y)-sin(z)| sin(x)-sin(z)| cos(x)+cos(y)>):
> curlF:=Curl(F);
```

$$curlF := [$$
$$(x, y, z) \to -\sin(y) + \cos(z), (x, y, z) \to -\cos(z) + \sin(x), (x, y, z) \to \cos(x) - \cos(y)$$
$$]$$

Since the curl is non-zero, we can replace the integral along C by a integral along the line segment L from $(0,0,0)$ to $(\pi,\pi,0)$ plus an integral of the curl over the rectangle S with vertices $(0,0,0)$, $(0,0,\pi)$, (π,π,π) and $(\pi,\pi,0)$. The parametrized line and the velocity are the same as above. The restricted vector field and the line integral are:

```
> Fr:=F&@(r&@t);
```

$$Fr := [\sin(t), \sin(t), 2\cos(t)]$$

```
> Int(Fr &. v, t=0..Pi); I_L:=value(%);
```

$$\int_0^\pi 2\sin(t)\,dt$$

$$I\_L := 4$$

The parametrized rectangle and the tangent and normal vectors are:

```
> R:=MF([s,t], <t|t|s>);
```

$$R := [(s, t) \to t, (s, t) \to t, (s, t) \to s]$$

```
> Rs:=D[1](R) &@ <t|t|s>; Rt:=D[2](R) &@ <t|t|s>; N:=Rs &x Rt;
```

$$Rs := [0, 0, 1]$$

$$Rt := [1, 1, 0]$$

$$N := [-1, 1, 0]$$

Applying the right hand rule to the normal, we check that $\partial S = C - L$ as it should be. The restriction of the curl to the surface is

```
> curlFR:=curlF&@(R &@ <s,t>);
```

$$curlFR := [-\sin(t) + \cos(s),\ -\cos(s) + \sin(t),\ 0]$$

and the surface integral is

```
> Muint(curlFR &.  N, s=0..Pi, t=0..Pi); I_S:=value(%);
```

$$\int_0^{\pi} \int_0^{\pi} -2\cos(s) + 2\sin(t)\, ds\, dt$$

$$I\_S := 4\,\pi$$

So the line integral along C is

```
> I_C:=I_L + I_S;
```

$$I\_C := 4 + 4\,\pi$$

Example: 8.17. Compute the line integral $\oint_C \vec{F} \cdot d\vec{s}$ for the vector field $\vec{F} = \left(\frac{-y}{x^2+y^2}, \frac{x}{x^2+y^2}, 0 \right)$ counterclockwise around the closed curve C which is the octagon with vertices $(3,-2,3)$, $(3,2,3)$, $(2,3,2)$, $(-2,3,-2)$, $(-3,2,-3)$, $(-3,-2,-3)$, $(-2,-3,-2)$ and $(2,-3,2)$.

SOLUTION: We first enter the vector field and compute the curl:

```
> F:=MF([x,y,z], <-y/(x^2+y^2)| x/(x^2+y^2)| 0>):
> Curl(F);
```

$$[(x, y, z) \to 0,\ (x, y, z) \to 0,\ (x, y, z) \to 0]$$

Thus the curl is zero. Normally, we would use Stokes' Theorem and conclude the line integral is zero. However, notice that the vector field is not defined on the z-axis. So the curl is also undefined on the z-axis and Stokes' Theorem does not apply. Thus the octagon may be replaced by any closed curve which also circles the z-axis once counterclockwise. The simplest such curve is the circle $x^2 + y^2 = 1$ in the xy-plane:

```
> r:=MF(t, <cos(t)|sin(t)|0>):
```

The velocity, vector field, $\vec{F}$, and the line integral are:

```
> v:=D(r) &@ t;
```

$$v := [-\sin(t),\ \cos(t),\ 0]$$

```
> Fr:=simplify(F&@(r &@ t));
```

$$Fr := [-\sin(t),\ \cos(t),\ 0]$$

```
> Int(Fr &.  v, t=0..2*Pi); value(%);
```

$$\int_0^{2\,\pi} 1\, dt$$

$$2\,\pi$$

8.6.2 Related Surface and Volume Integrals

Suppose S_1 and S_2 are two *open* surfaces which stay in a region R and have the same boundary curve, $\partial S_1 = \partial S_2$. We say that S_1 can be continuously deformed into S_2 within R if there is a solid region V within R whose boundary is $\partial V = S_2 - S_1$. This means that the boundary of V with outward normal consists of S_2 with its normal unchanged and S_1 with its normal reversed.

Similarly, suppose S_1 and S_2 are two *closed* surfaces which stay in a region R. We say that S_1 can be continuously deformed into S_2 within R if there is a solid region V within R whose boundary is $\partial V = S_2 - S_1$. This means that the boundary of V with normal pointing out of V consists of the two surfaces S_1 and S_2 with the normal of S_1 reversed.

In either case, $\partial V = S_2 - S_1$ and so Gauss' Theorem says

$$\iiint\limits_V \vec{\nabla} \cdot \vec{F}\, dV = \iint\limits_{\partial V} \vec{F} \cdot d\vec{S} = \iint\limits_{S_2} \vec{F} \cdot d\vec{S} - \iint\limits_{S_1} \vec{F} \cdot d\vec{S}$$

This may be rewritten as

$$\iint\limits_{S_2} \vec{F} \cdot d\vec{S} = \iint\limits_{S_1} \vec{F} \cdot d\vec{S} + \iiint\limits_V \vec{\nabla} \cdot \vec{F}\, dV$$

In other words, to compute $\iint\limits_{S_2} \vec{F} \cdot d\vec{S}$, you may alternatively compute $\iint\limits_{S_1} \vec{F} \cdot d\vec{S}$ and $\iiint\limits_V \vec{\nabla} \cdot \vec{F}\, dV$, if that is easier.

In the special case that $\vec{\nabla} \cdot \vec{F} = 0$ everywhere in the region R, then

$$\iint\limits_{S_2} \vec{F} \cdot d\vec{S} = \iint\limits_{S_1} \vec{F} \cdot d\vec{S}$$

whenever S_1 can be continuously deformed into S_2 within R. For the case of open surfaces, this is a special case of the surface independence discussed at the beginning of section 8.4.2.

Example: 8.18. For each of the following velocity fields, compute the flux integral $\iint\limits_C \vec{v} \cdot d\vec{S}$ through the cone C given by $z = \sqrt{x^2 + y^2}$ for $0 \le z \le 2$ with the normal pointing in and up.

a) $v = (x, y, -2z)$.

b) $v = (x, y, -3z)$.

SOLUTION:

a) We first enter the velocity field and compute the divergence:

```
> F:=MF([x,y,z], <x| y| -2*z>):
> Div(F);
```

$$(x,\, y,\, z) \to 0$$

Since the divergence is zero, we can replace the integral over the cone C by a integral over the disk D given by $x^2 + y^2 \leq 4$ with $z = 2$. The parametrized disk and the flux integral are

```
> R:=MF([r,theta],<r*cos(theta)| r*sin(theta)| 2>):
> Siv(F,R, r=0..2, theta=0..2*Pi); value(%);
```

$$\int_0^{2\pi} \int_0^2 -4\,r\,dr\,d\theta$$

$$-16\,\pi$$

b) We again enter the velocity field and compute the divergence:

```
> F:=MF([x,y,z], <x| y| -3*z>):
> divF:=Div(F)(x,y,z);
```

$$divF := -1$$

Since the divergence is non-zero, we can replace the integral over the cone C by a integral over the disk D given by $x^2 + y^2 \leq 4$ with $z = 2$ *minus* the integral of the divergence over the solid cone between $z = \sqrt{x^2 + y^2}$ and $z = 2$. The volume integral is subtracted because the normal to the cone points into the volume. The parametrized disk is given above. So the surface integral is

```
> Siv(F,R, r=0..2, theta=0..2*Pi); I_D:=value(%);
```

$$\int_0^{2\pi} \int_0^2 -6\,r\,dr\,d\theta$$

$$I\_D := -24\,\pi$$

The solid cone may be parametrized in cylindrical coordinates. So the integral of the divergence is

```
> Muint(divF*r, z=r..2, r=0..2, theta=0..2*Pi); I_V:=value(%);
```

$$\int_0^{2\pi} \int_0^2 \int_r^2 -r\,dz\,dr\,d\theta$$

$$I\_V := -\frac{8\,\pi}{3}$$

Notice that this is just the negative of the volume of a cone of radius 2 and height 2. So the flux integral through C is

```
> I_C:=I_D - I_V;
```

$$I\_C := -\frac{64\,\pi}{3}$$

Example: 8.19. For the vector field $\vec{F} = \left(\frac{x}{(x^2 + y^2 + z^2)^{3/2}}, \frac{y}{(x^2 + y^2 + z^2)^{3/2}}, \frac{z}{(x^2 + y^2 + z^2)^{3/2}} \right)$, compute the surface integral $\iint_S \vec{F} \cdot d\vec{S}$ over the closed surface S of the rectangular solid $-5 \leq x \leq 5$, $-4 \leq y \leq 4$, $-3 \leq z \leq 3$ with outward normal.

SOLUTION: We first enter the vector field and compute the divergence:

```
> F:=MF([x,y,z], <x/(x^2+y^2+z^2)^(3/2)| y/(x^2+y^2+z^2)^(3/2)|
> z/(x^2+y^2+z^2)^(3/2)>):
```

```
> divF:=Div(F)(x,y,z);
```

$$divF := 0$$

Thus the divergence is zero. Normally, we would use Gauss' Theorem and conclude the surface integral is zero. However, notice that the vector field is not defined at the origin. So the divergence is also undefined at the origin and Gauss' Theorem does not apply. Thus the surface of the rectangular solid may be replaced by any closed surface which also encloses the origin. The simplest such surface is the sphere $x^2 + y^2 + z^2 = 1$:

```
> R:=MF([phi,theta], <sin(phi)*cos(theta)| sin(phi)*sin(theta)|
> cos(phi)>):
```

(Notice we have ordered the parameters (ϕ, θ) so the normal points out.) So the surface integral is:

```
> Siv(F, R, theta=0..2*Pi, phi=0..Pi); value(%);
```

$$\int_0^{\pi} \int_0^{2\pi} \sin(\phi)\, d\theta\, d\phi$$

$$4\pi$$

8.7 Exercises

- Do Labs: 9.9, 9.10, 9.11 and 9.12.
- Do Project: 10.14.

1. Verify the Fundamental Theorem of Calculus for Curves by computing both sides for the indicated function and curve:

   (a) $f = x + y^2 + z^3$ and $\vec{r}(t) = (t^3, t^2, t)$ for $1 \le t \le 2$.

   (b) $f = xy^2z^3w^4$ and $\vec{r}(t) = (t^4, t^3, t^2, t)$ for $1 \le t \le 2$.

2. Find a scalar potential f for the vector field $\vec{F}(x, y) = (2xy^3, 3x^2y^2)$. Then use the Fundamental Theorem of Calculus for Curves to evaluate $\int_C \vec{F} \cdot d\vec{s}$, where C is the line segment from $A = (2,3)$ to $B = (5,2)$. Verify your result by computing the line integral directly.

3. Show that the line integral $\int_C (2x \sin y)\, dx + (x^2 \cos y - 3y^2)\, dy$ is independent of path and evaluate it on any curve between $(-1, 0)$ and $(5, 1)$ using the Fundamental Theorem of Calculus for Curves.

4. Find a scalar potential f for the vector field $\vec{F} = (y, x)$ and use it to evaluate $\int_C \vec{F} \cdot d\vec{s}$, where C is the arc of the quartic curve $y = x^4 - x^3$ from $(1, 0)$ to $(2, 8)$.

5. Compute each of the following integrals along the spiral $\vec{r}(t) = (t\cos t, t\sin t)$ for $0 \le t \le \dfrac{7\pi}{3}$. If the integral is path independent, you may find a potential and use the Fundamental Theorem of Calculus for Curves.

   (a) $\displaystyle\int x\,dx + y\,dy$ (b) $\displaystyle\int x\,dx - y\,dy$ (c) $\displaystyle\int y\,dx - x\,dy$ (d) $\displaystyle\int y\,dx + x\,dy$

6. Compute each of the following integrals once counterclockwise around the ellipse $\vec{r}(\phi) = (4\cos\phi, 3\sin\phi)$. If the integral is path independent, you may find a potential and use the Fundamental Theorem of Calculus for Curves.

   (a) $\displaystyle\oint x\,dx + y\,dy$ (b) $\displaystyle\oint x\,dx - y\,dy$ (c) $\displaystyle\oint y\,dx - x\,dy$ (d) $\displaystyle\oint y\,dx + x\,dy$

7. Verify Green's Theorem by computing both sides for the line integral $\displaystyle\oint_C x^4y^5\,dx + x^7y^6\,dy$, where C is the circle $x^2 + y^2 = 4$ traversed once counterclockwise.

8. Use Green's Theorem to evaluate the line integral $\displaystyle\oint_C (y + e^{\sqrt{x}})\,dx + (2x + \cos(y^2))\,dy$, where C is the boundary of the region enclosed by the parabolas $y = x^2$ and $x = y^2$ traversed counterclockwise.

9. Repeat exercise 6 but use Green's Theorem to do the integrals.

10. Compute the area of the region inside the general ellipse $\dfrac{x^2}{a^2} + \dfrac{y^2}{b^2} = 1$.

    HINT: The general ellipse may be parametrized by $\vec{r}(\theta) = (a\cos\theta, b\sin\theta)$.

11. Consider a propeller with three blades, the front face of which has the parametric boundary $\vec{r}(t) = \big((3 + 2\cos 3t)\cos t, (3 + 2\cos 3t)\sin t\big)$, for $0 \le t \le 2\pi$. Use Green's Theorem to compute the area of the face of the propeller.

12. Verify Stokes' Theorem by computing both sides for the vector field $\vec{F} = (3x^2y, -3xy^2, 0)$ and the paraboloid $z = x^2 + y^2$ for $z \le 4$ with the normal pointing up and in.

13. Verify Stokes' Theorem by computing both sides for the vector field $\vec{F} = (yz, y^2, -xy)$ and the surface S which is the elliptic paraboloid $y = x^2 + z^2$ for $y \le 9$ with normal pointing in and up along the y-axis.

14. Use Stokes' Theorem to evaluate the surface integral $\displaystyle\iint_S \vec{\nabla} \times \vec{F} \cdot d\vec{S}$, where the surface S is the part of the paraboloid $y = 1 - x^2 - z^2$ that lies to the right of the xz-plane, oriented toward the xz-plane, and the vector field is $\vec{F}(x, y, z) = (yz^3, \sin(xyz), x^3)$.

15. Let T be the triangular surface with vertices $P = (0, 0, 2)$, $Q = (2, 2, 2)$ and $R = (2, 0, 2)$ and let C be its boundary path traversed from P to Q to R to P. Use *rectangular* coordinates to parametrize T; i.e., $\vec{R}(x, y) = (\_\_, \_\_, \_\_)$. The shadow region (the projection of the triangular surface onto the xy-plane where $z = 0$) is the triangular region with vertices $P' = (0, 0)$, $Q' = (2, 2)$ and $R' = (2, 0)$. Accordingly, what are the

ranges of x and y? Then use Stokes' Theorem to compute the line integral $\oint_C \vec{F}\cdot d\vec{s}$ where $\vec{F}(x,y,z) = (x^2y^2, x^2z^2, y^2z^2)$.

16. Use Stokes' Theorem to evaluate the line integral $\oint_C \vec{F}\cdot d\vec{s}$ where $\vec{F}(x,y,z) = (x^2z, xy^2, z^2)$ and C is the curve of intersection of the plane $x+y+z=1$ with the cylinder $x^2+y^2=9$, oriented counterclockwise as viewed from above.

17. Verify Gauss' Theorem by computing both sides for the indicated vector field and solid region:

    (a) $F = (x^3, y^3, z^3)$ and V is the region inside of the sphere $x^2+y^2+z^2=4$.

    (b) $F = \big(\cos(x), \cos(y), \cos(z)\big)$ and V is the region inside of the cube $0 \le x \le \pi$, $0 \le y \le \pi$ and $0 \le z \le \pi$.

18. Use the Gauss' Theorem to calculate the surface integral $\iint_S \vec{F}\cdot d\vec{S}$ for the vector field $\vec{F}(x,y,z) = (x^3, y^3, z^3)$, where S is surface of the solid bounded by the cylinder $x^2+y^2=1$ and the planes $z=0$ and $z=2$ with outward normal.

19. Use the Gauss' Theorem to compute the surface integral $\iint_S \vec{F}\cdot d\vec{S}$, where S is the sphere $x^2+y^2+z^2=9$ oriented outward and $\vec{F}(x,y,z) = (2xz^2, -yx^2, 3zy^2)$.

20. Do a surface integral to compute the volume of the region inside the general ellipsoid $\dfrac{x^2}{a^2}+\dfrac{y^2}{b^2}+\dfrac{z^2}{c^2}=1$.

    HINT: The general ellipsoid may be parametrized by $\vec{R}(\theta,\phi) = (a\sin\phi\cos\theta, b\sin\phi\sin\theta, c\cos\phi)$.

21. For each of the following vector fields, compute the line integral $\int_P \vec{F}\cdot d\vec{s}$ along the parabola $y = z = 16 - x^2$ from $(-4,0,0)$ to $(4,0,0)$. If the line integral is path independent, you should change the path to a straight line. If not, you should apply Stokes' Theorem to the surface between the parabola and the line.
    (a) $\vec{F} = (e^x, e^y, e^z)$ (b) $\vec{F} = (yz, -xz, xy)$

22. For each of the following vector fields, compute the line integral $\oint_T \vec{F}\cdot d\vec{s}$ around the closed equilateral triangle T from $(0,2,3)$ to $\left(\sqrt{3},-1,3\right)$ to $\left(-\sqrt{3},-1,3\right)$ and back to $(0,2,3)$. If the line integral is path independent, you should change the path to the circle $x^2+y^2=4$ in the plane $z=3$. If not, you should apply Stokes' Theorem to the surface between the triangle and the circle.

    (a) $\vec{F} = \left(-\dfrac{y}{x^2+y^2}, \dfrac{x}{x^2+y^2}, \dfrac{1}{z^2}\right)$ (b) $\vec{F} = \left(\dfrac{1}{x^2+y^2}, \dfrac{1}{x^2+y^2}, 0\right)$

23. For each of the following vector fields, compute the surface integral $\iint_Q \vec{F} \cdot d\vec{S}$ over the quartic surface $z = (16 - x^2)(9 - y^2) + 5$ above the rectangle $-4 \le x \le 4$ and $-3 \le y \le 3$ with upward normal. If the surface integral is surface independent, you should change the surface to a rectangle in the plane $z = 5$. If not, you should apply Gauss' Theorem to the volume between the quartic and the rectangle.
(a) $\vec{F} = (y^4, z^4, x^4)$ (b) $\vec{F} = (x^4, y^4, z^4)$

24. For each of the following vector fields, compute the surface integral $\iint_S \vec{F} \cdot d\vec{S}$ over the total closed surface of the cylinder $x^2 + y^2 \le 9$ and $-4 \le z \le 4$ with outward normal. If the surface integral is surface independent, you should change the surface to the sphere $x^2 + y^2 + z^2 = 25$. If not, you should apply Gauss' Theorem to the volume between the cylinder and the sphere.
(a) $\vec{F} = \left(\frac{x}{(x^2 + y^2 + z^2)^{3/2}}, \frac{y}{(x^2 + y^2 + z^2)^{3/2}}, \frac{z}{(x^2 + y^2 + z^2)^{3/2}} \right)$
(b) $\vec{F} = \left(\frac{x}{x^2 + y^2 + z^2}, \frac{y}{x^2 + y^2 + z^2}, \frac{z}{x^2 + y^2 + z^2} \right)$

Chapter 9

Labs

This chapter contains a collection of labs on vector calculus. Typically, the students should work in pairs in class for about 1 hour. If they don't finish, they should complete the lab on their own time during the week and turn it in at the next week's lab. A short lab report is expected. The report should be graded on mathematics, *Maple* and English presentation.

- 9.1 Orienteering
- 9.2 Dot and Cross Products
- 9.3 Lines, Planes, Quadric Curves and Quadric Surfaces
- 9.4 Frenet Analysis of Curves
- 9.5 Parametric Curves
- 9.6 Linear and Quadratic Approximations
- 9.7 Multivariable Max-Min Problems
- 9.8 A Volume of Desserts
- 9.9 Gauss' Law
- 9.10 Ampere's Law
- 9.11 Interpretation of the Divergence
- 9.12 Interpretation of the Curl

9.1 Lab: Orienteering

Objectives: You will learn to use *Maple* to perform vector addition and scalar multiplication, to convert between rectangular and polar or spherical coordinates and to plot points and dot-to-dot pictures.

You are strongly encouraged to work with a partner.

Before Lab: Read subsections 1.1.1 and 1.1.2 and section 1.2. Also read the *Maple* help pages:

```
> ?plot

> ?plot,options

> ?spacecurve

> ?plot3d,options
```

and the `VecCalc` help page:

```
> ?simplifyvec
```

Maple Commands: You will need to use the *Maple* commands for addition and scalar multiplication of vectors, and the following `plot` and `spacecurve` commands:

- *Maple* can `plot` a list of points as follows:

```
> plot([[1,0], [2,3], [3,0], [0,2], [4,2]], style=point,
> symbol=diamond, symbolsize=24);
```

Look at the help on `plot,options` to turn off the axes or change the `symbol` or `symbolsize`.

- If you leave off the option `style=point`, *Maple* will connect the dots with line segments. To connect back to the start, you must repeat the starting point:

```
> plot([[1,0], [2,3], [3,0], [0,2], [4,2], [1,0]], axes=none);
```

What shape did you get?

- You can also plot points and dot-to-dot pictures in 3-dimensions. For example, here is a cube:

```
> spacecurve({[[0,0,0], [0,1,0], [1,1,0], [1,0,0], [0,0,0], [0,0,1],
> [0,1,1], [1,1,1], [1,0,1], [0,0,1]], [[0,1,0], [0,1,1]], [[1,1,0],
> [1,1,1]], [[1,0,0], [1,0,1]]}, orientation=[30,60]);
```

Notice that several dot-to-dot pieces are put together by enclosing them in braces and separating them by commas. Try dragging the plot with you mouse.

`VecCalc` Commands: Enter your points as ordered pairs or triples in square brackets, `[ ]`, not angle brackets, `< >`. You may need to use the `VecCalc` commands `r2d` (convert radians to degrees), `d2r` (convert degrees to radians), `r2p` (convert rectangular to polar), `p2r` (convert polar to rectangular), `r2s` (convert rectangular to spherical), `s2r` (convert spherical to rectangular).

Lab Requirements: Start the `VecCalc` package by executing:

```
> with(VecCalc): VCalias:
```

Answer the following questions. Number each problem either in a text region or using a *Maple* comment. Where appropriate, you must explain your reasoning in text regions. Save your file now and after each problem. Print your worksheet to turn in as a lab report.

1. **Orienteering:** You start at the origin and travel South-East for 15 paces. Then you travel West-South-West for 20 paces. Finally you travel North-North-West for 27 paces. Construct a vector for each of these travel segments. If you want to go directly back to the origin, in what direction should you travel and how many paces will it take? Give the direction in degrees East or West of North. Plot your path. Remember, South-South-East is halfway between South and South-East. Also remember, *Maple* works in radians!

2. **Going Hunting:** Plot the constellation Orion as shown below:

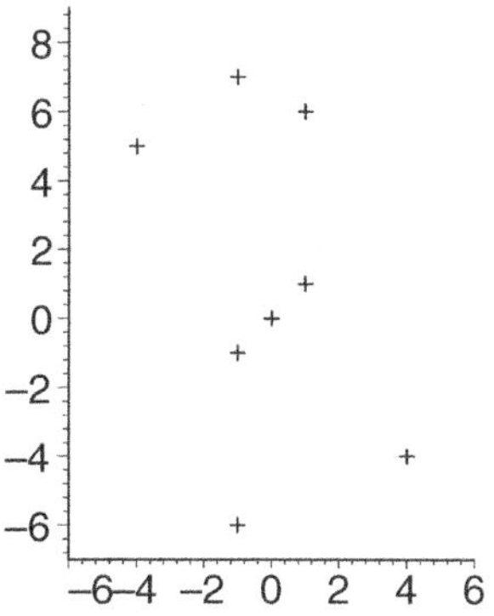

3. **Starfleet 3D Orienteering:** Galactic Coordinates are specified by taking the origin at the center of mass of the galaxy, with the galaxy in the xy-plane, the x-axis passing through the sun, (We're still heliocentric!) and the z-axis specified by the right hand rule so that when you are on the positive z-axis, the galaxy rotates counterclockwise from the positive x-axis to the positive y-axis.

   You start at the galactic origin and successively make each of the following motions. Where do you end up? Plot your path. Remember, *Maple* works in radians!

   Each motion is specified in spherical coordinates where ρ is the distance you travel, ϕ is the polar angle measured down from the positive z-axis and θ is the azimuthal angle measured counterclockwise from the positive x-axis. Give your final position in spherical coordinates.

   (a) $(\rho, \theta, \phi) = (2 \text{ lightyears}, 30°, 135°)$

   (b) $(\rho, \theta, \phi) = (4 \text{ lightyears}, 300°, 120°)$

   (c) $(\rho, \theta, \phi) = (3 \text{ lightyears}, 225°, 30°)$

9.2 Lab: Dot and Cross Products

Objectives: You will learn the *Maple* commands for 3-dimensional analytic geometry and vectors.

You are strongly encouraged to work with a partner.

Before Lab: Read sections 1.1 and 1.2. Do problem 1 below by hand; you will redo it in lab using *Maple*. You are expected to turn in this hand computation before lab.

Maple Commands: You will need to use *Maple*'s assignment statements and arithmetic operators and the *Maple* commands `expand`, `evalf` and `spacecurve`. See lab 9.1 for an example using `spacecurve`.

`VecCalc` Commands: `Dot` (dot product), `Length` (length of a vector), `Cross` (cross product) and `r2d` (convert radians to degrees).

Lab Requirements: Start the `VecCalc` package by executing:

```
> with(VecCalc): VCalias:
```

Answer the following questions. Number each problem either in a text region or using a *Maple* comment. Where appropriate, you must explain your reasoning in text regions. Save your file now and after each problem. Print your worksheet to turn in as a lab report.

1. Derive the identity
$$(\vec{u}\cdot\vec{v})^2 + |\vec{u}\times\vec{v}|^2 = |\vec{u}|^2|\vec{v}|^2$$
as follows: Let $\vec{u} = (u_1, u_2, u_3)$ and $\vec{v} = (v_1, v_2, v_3)$.
   (a) Write out $(\vec{u}\cdot\vec{v})^2$ to get 6 terms.
   (b) Write out $|\vec{u}\times\vec{v}|^2$ to get 9 terms.
   (c) Add $(\vec{u}\cdot\vec{v})^2 + |\vec{u}\times\vec{v}|^2$ and cancel some terms.
   (d) Multiply out $|\vec{u}|^2|\vec{v}|^2$.
   (e) Check the answers to (c) and (d) are equal by subtracting them.

In problems 2 – 4, let $\vec{u} = (2, -4, 1)$ and $\vec{v} = (3, -2, -1)$.

2. Find the angle between the vectors $\vec{u}$ and $\vec{v}$ in degrees.
3. Find the scalar and vector projections of $\vec{v}$ along $\vec{u}$.
4. Find the area of the triangle with edges $\vec{u}$ and $\vec{v}$.

---

5. Given the points $P = (3, -1, 4)$, $Q = (-2, 4, 3)$, $R = (2, 1, 5)$, and $S = (3, -5, 0)$, find the volume of the parallelepiped with adjacent edges $\overrightarrow{PQ}$, $\overrightarrow{PR}$, and $\overrightarrow{PS}$. Then find the other four vertices (e.g. $R + \overrightarrow{PQ}$) and plot the parallelepiped using `spacecurve`.

6. Show that the three points $A = (4, 2, 3)$, $B = (2, 2, 5)$ and $C = (4, 0, 5)$ are the vertices of an equilateral triangle by computing the three angles and the lengths of the three edges. Plot the triangle using `spacecurve` and rotate the plot so you can see it is equilateral.

9.3 Lab: Lines, Planes, Quadric Curves and Quadric Surfaces

Objectives: You will learn to use *Maple* to solve problems involving lines, planes, quadric curves and quadric surfaces.

You are strongly encouraged to work with a partner.

Before Lab: Read sections 1.1 and 1.3.

Maple Commands: `solve`, `completesquare`, `implicitplot` and `implicitplot3d`.

`VecCalc` **Commands:** `MF` or `&->` (make a function), `EF` or `&@` (evaluate a function), `Dot` or `&.` (dot product), `Length` (length of a vector), `Angle` (angle between vectors), `Cross` or `&x` (cross product) and `r2d` (convert radians to degrees).

Lab Requirements: Start the `VecCalc` package by executing:

```
> with(VecCalc): VCalias:
```

Answer the following questions. Number each problem either in a text region or using a *Maple* comment. Where appropriate, you must explain your reasoning in text regions. Save your file now and after each problem. Print your worksheet to turn in as a lab report.

1. Consider the line $(x, y, z) = (2 - 5t, -1 + t, 7 - 3t)$ and the plane $4x - 2y + 3z = 62$. Find the angle (correct to the nearest degree) between the line and the normal to the plane and determine if the line and plane are parallel or perpendicular or neither. If they are not parallel, find their point of intersection.

2. Show that the planes $3x + y - 2z = 3$ and $x + 2y + 3z = 1$ are neither parallel nor perpendicular by finding (correct to the nearest degree) the angle between their normals. Then find parametric equations for their line of intersection. (HINT: Let $z = t$ and solve these three equations for x, y and z.)

---

3. The points $A = (4, 2, 3)$, $B = (2, 2, 5)$ and $C = (4, 0, 5)$ are the vertices of an equilateral triangle. (See lab 9.2.) Find the center D of the triangle $\triangle ABC$ by finding the intersection of two lines, each from a vertex to the midpoint of the opposite side. Then find the equation of the line perpendicular to the triangle passing through the point D.

---

4. Follow the directions of example 1.14 for the following quadric curves:

   (a) $x^2 - 3y^2 - 6x + 12y = 6$

   (b) $x^2 + 3y^2 - 6x + 12y = 6$

   (c) $x^2 + 3y^2 - 6x + 12y = -25$

   (d) $3y^2 - 6x + 12y = -24$

5. Follow the directions of example 1.15 for the following quadric surfaces:

   (a) $x^2 - 3y^2 + 4z^2 - 6x + 6y = -7$

   (b) $x^2 - 3y^2 - 4z^2 - 6x + 6y = -7$

   (c) $x^2 + 3y^2 + 4z^2 - 6x + 6y = -3$

   (d) $x^2 - 3y^2 - 6x + 6y = -3$

   (e) $x^2 - 3y^2 - 4z^2 - 6x + 6y = -6$

   (f) $-3y^2 + 4z^2 - 5x + 6y = -7$

   (g) $3y^2 + 4z^2 - 5x + 6y = 7$

9.4 Lab: Frenet Analysis of Curves

Objectives: You will learn the *Maple* commands for the geometric properties of space curves.

You are strongly encouraged to work with a partner.

Before Lab: Read section 2.2 and the `VecCalc` help page

```
> ?Curve
```

Also recall: $\sinh(x) = \dfrac{e^x - e^{-x}}{2}$ $\quad \cosh(x) = \dfrac{e^x + e^{-x}}{2}$

Maple Commands: You will need to use *Maple*'s trigonometric and hyperbolic functions and the *Maple* commands `spacecurve`, `Diff`, `Int`, `value` and `simplify`.

`VecCalc` Commands: `Dot` or `&.` (dot product), `Length` (length of a vector), `Cross` or `&x` (cross product), `Cv` (Curve velocity), `Ca` (Curve acceleration), `Cj` (Curve jerk), `Cs` (Curve speed), `CL` (Curve arc length), `CT` (Curve unit tangent), `CB` (Curve unit bonormal), `CN` (Curve unit principal normal), `Ck` (Curve curvature), `Ct` (Curve torsion), `CaT` (Curve tangential acceleration), and `CaN` (Curve normal acceleration). Use `simplifyvec` if *Maple* does not simplify scalar multiplication.

Lab Requirements: Consider one of the following space curves. (Your instructor will individually tell you which to use.)

$$\vec{r}(t) = (\ 3\cosh(t),\ 3\sinh(t),\ 3t\ ) \quad \text{or} \quad \vec{r}(t) = (\ 2t\cos(t),\ 2t\sin(t),\ 2t\ )$$

$$\text{or} \quad \vec{r}(t) = (\ t^2\cos(t),\ t^2\sin(t),\ 2t\ ) \quad \text{or} \quad \vec{r}(t) = (\ t+\ln(t),\ t-\ln(t),\ 2t\ )$$

Compute the items below with *Maple*. If necessary, use `simplify` or `simplifyvec` to clean up final expressions. Among the `VecCalc` commands, you may **only** use the commands `Dot`, `Length` and `Cross` to compute the quantities. The mathematical definitions of the various quantities are provided to aid you in this semiautomatic computation. You should then use the `VecCalc` Curve commands to check your work in a fully automatic fashion.

Answer the following questions. Number each problem either in a text region or using a *Maple* comment. Where appropriate, you must explain your reasoning in text regions. Save your file now and after each problem. Print your worksheet to turn in as a lab report.

1. Define $\vec{r}$ as a vector of expressions.
2. Plot $\vec{r}$ for $0 \le t \le 2$ using `spacecurve`.
3. Compute the velocity $\vec{v}$ using `Diff` and `value`. Check using `Cv`. $\quad \vec{v} = \vec{r}\,'$
4. Compute the acceleration $\vec{a}$. Check using `Ca`. $\quad \vec{a} = \vec{v}\,' = \vec{r}\,''$
5. Compute the jerk $\vec{j}$. Check using `Cj`. $\quad \vec{j} = \vec{a}\,' = \vec{v}\,'' = \vec{r}\,'''$
6. Compute the speed $\dfrac{ds}{dt}$ using `Length`. Check using `Cs`. $\quad \dfrac{ds}{dt} = |\vec{v}|$
7. Compute the arc length L for $0 \le t \le 2$ using `Int` and `value`.

   Check using `CL`. $\quad L = \displaystyle\int_0^2 |\vec{v}|\,dt$
8. Compute the unit tangent vector $\hat{T}$. Check using `CT`. $\quad \hat{T} = \dfrac{\vec{v}(t)}{|\vec{v}(t)|}$
9. Compute the unit binormal vector $\hat{B}$ using `Cross` and `Length`.

   Check using `CB`. $\quad \hat{B} = \dfrac{\vec{v} \times \vec{a}}{|\vec{v} \times \vec{a}|}$
10. Compute the unit principal normal vector $\hat{N}$. Check using `CN`. $\quad \hat{N} = \hat{B} \times \hat{T}$
11. Compute the curvature κ. Check using `Ck`. $\quad \kappa = \dfrac{|\vec{v} \times \vec{a}|}{|\vec{v}|^3}$
12. Compute the torsion τ. Check using `Ct`. $\quad \tau = \dfrac{\vec{v}(t) \times \vec{a}(t) \cdot \vec{j}(t)}{|\vec{v}(t) \times \vec{a}(t)|^2}$
13. Compute the tangential acceleration a_T (two ways). Check using `CaT`. $\quad a_T = \vec{a} \cdot \hat{T} = \dfrac{d}{dt}|\vec{v}(t)|$
14. Compute the normal acceleration a_N (two ways). Check using `CaN`. $\quad a_N = \vec{a} \cdot \hat{N} = \kappa(t)|\vec{v}(t)|^2$

9.5 Lab: Parametric Curves

Objectives: You will learn to use *Maple* to plot parametric curves, to find intersections of parametric curves with various lines and to find slopes and self-intersections of parametric curves.

You are strongly encouraged to work with a partner.

Before Lab: Read sections 1.3 and 2.1. Also read the *Maple* help pages:

```
> ?plot,parametric
> ?textplot
> ?fsolve
```

In particular, look at the example of `fsolve` with two equations, two variables and two intervals.

Maple Commands:

- The following commands plot the parametric curve $x = \cos(t), y = \sin(t)$ for $0 \leq t \leq 2\pi$.

```
> x0:= t-> cos(t); y0:= t-> sin(t);
> plot([x0(t), y0(t), t=0..2*Pi]);
```

Notice the square brackets; they're necessary and the parameter range must be inside the brackets. Also notice that we named the coordinates `x0` and `y0` rather than `x` and `y` so that `x` and `y` would still be available to be used in equations.

- The following commands plot a point and label its coordinates.

```
> pp:=plot([[2,1]], style=point, symbol=cross, symbolsize=36):
> tp:=textplot([2,1, "  (2,1)"], align={ABOVE,RIGHT}):
> display([pp,tp], view=[0..3,0..2]);
```

- To solve an equation, it is often best to use `fsolve` in conjunction with a plot. For example, to solve the equation: $2\sin(x) = x + \cos(x)$ you should first plot the two functions:

```
> f:=x->2*sin(x); g:=x->x+cos(x);
> plot({f(x),g(x)}, x=-5..5);
```

Notice there are three solutions in the intervals $[-2,-1]$, $[.5,1.5]$ and $[2,3]$. So you can now use `fsolve` with intervals:

```
> fsolve(f(x)=g(x),x=-2..-1);
> fsolve(f(x)=g(x),x=.5..1.5);
> fsolve(f(x)=g(x),x=2..3);
```

- To solve two equation for two unknowns, use `fsolve` with the following syntax: For example, to solve the equations $s - 3t = 0$ and $st = 5$, execute:

```
> fsolve({s-3*t=0, s*t=5}, {s=3..6,t=1..2});
```

Lab Requirements: [1]Start the plots package by executing:

```
> with(plots):
```

Answer the following questions. Number each problem either in a text region or using a *Maple* comment. Where appropriate, you must explain your reasoning in text regions. Save your file now and after each problem. Print your worksheet to turn in as a lab report.

This lab concerns the parametric curve

$$x = 3\sin(2\pi t) - 2\cos^5(2\pi t) \quad y = \cos(2\pi t) - 3\sin(2\pi t) \qquad \text{for} \quad 0 \le t \le 1 .$$

1. Plot the curve. Next, plot the points on the curve where t is 0, 0.1, 0.2, 0.3, 0.4, 0.5, 0.6, 0.7, 0.8, 0.9, and 1. See lab 9.1 for how to plot points. Then label the points using **textplot**. Combine the three plots using the **display** command in the **plots** package .

2. Determine the points (x, y) where the curve crosses the line $x = 1$.

   HINT: You'll have to find the values of t which make the x coordinate equal to 1 by using **fsolve**. A plot of $x(t)$ may be helpful. Then plug t into x and y.

3. Determine the points (x, y) where the curve crosses the line $y = 2x$.

   HINT: You'll have to find the t values which solve $y(t) = 2x(t)$.

4. Find the points (x, y) where the tangent line to the curve is horizontal.

   HINT: $\dfrac{dy}{dx} = \dfrac{dy/dt}{dx/dt}$, so you'll need to solve $\dfrac{dy}{dt} = 0$ for t.

5. Find the points (x, y) where the tangent line to the curve is vertical.

6. Find the points (x, y) where the tangent line has slope $\dfrac{-2}{3}$.

   HINT: It's easier to work with the equation $\dfrac{dy}{dt} = \dfrac{-2}{3}\dfrac{dx}{dt}$ than the equation $\dfrac{dy/dt}{dx/dt} = \dfrac{-2}{3}$.

7. Find the points (x, y) where the curve crosses itself.

   HINT: You'll need to find two different values, t_1 and t_2, so that $x(t_1) = x(t_2)$ and $y(t_1) = y(t_2)$. Do this by using **fsolve** on the pair of equations with ranges for t_1 and t_2.

8. (20% EXTRA CREDIT) At one point (x, y) where the curve crosses itself, find the angle in degrees between the two branches of the curve.

   HINT: Find the angle between the tangent vectors. Or find the slope and the inclination angle of each branch using **arctan** and subtract subtract inclinations. To see the angles properly in your plot, you must use the option **scaling=constrained**.

[1]This lab was suggested by Thomas Vogel, Texas A&M University.

9.6 Lab: Linear and Quadratic Approximations

Objectives: You will use *Maple* to find the linear and quadratic approximations to a surface in 3-dimensional space.

You are strongly encouraged to work with a partner.

Before Lab: Read subsections 3.2.1, 3.2.2, 3.2.3 and 3.2.7, especially examples 3.7 and 3.13.

Maple Commands: `diff`, `solve`, `fsolve`, `eval`, `simplify`, `evalf`, `implicitplot3d`, `plot3d`, `display` and `mtaylor`.

`VecCalc` **Commands:** `MF` or `&->` (Make Function), `EF` or `&@` (Evaluate Function), `Grad` (gradient) and `Dot` or `&.` (dot product).

Lab Requirements: Start the `VecCalc` package and define e:

```
> with(VecCalc): VCalias: e:=exp(1);
```

Consider the function

$$F(x, y, z) = x\ln(xz) + e^y \ln(yz)$$

and the surface S given by $F(x, y, z) = 3e$.

Answer the following questions. Number each problem either in a text region or using a *Maple* comment. Where appropriate, you must explain your reasoning in text regions. Save your file now and after each problem. Print your worksheet to turn in as a lab report.

1. Show that the point $P = (e, 1, e)$ lies on the surface S. Plot the surface S on the region $1 \le x \le 4$, $0.1 \le y \le 2$ and $1 \le z \le 4$ using `implicitplot3d` with the options `grid=[25,25,25], color=blue, axes=framed`.

2. The equation $F(x, y, z) = 3e$ implicitly defines z as a function of x and y, specifically $z = f(x, y)$. Use implicit differentiation to find $\dfrac{\partial f}{\partial x}$ and $\dfrac{\partial f}{\partial y}$ and their values at $(x, y) = (e, 1)$. See example 3.7.

3. Construct the linear approximation to $f(x, y)$ at $(x, y) = (e, 1)$. Then use it to estimate the value of $f(3, 1.02)$. Check it by using `fsolve` to find the exact value of $z = f(3, 1.02)$.

4. Plot the plane tangent to the surface S at the point $P = (e, 1, e)$ over the region $1 \le x \le 4$ and $0.1 \le y \le 2$ using `plot3d` with the options `view=1..4, grid=[25,25], color=red, axes=framed`. Then `display` it with the surface S using the option `orientation=[-10,70]`.

5. Recompute the equation of the tangent plane to the surface S at the point $P = (e, 1, e)$ by regarding S as a level surface of the function $F(x, y, z)$. See example 3.13.

6. Recompute the equation of the tangent plane to the surface S at the point $P = (e, 1, e)$ by computing the first order Taylor polynomial $p_1(x, y, z)$ for the function $F(x, y, z)$ at $(e, 1, e)$. See subsection 3.2.3.

7. Compute the second order Taylor polynomial $p_2(x, y, z)$ for the function $F(x, y, z)$ at $(e, 1, e)$. Approximate the surface S as the quadric $p_2(x, y, z) = 3e$. Then plot the quadric using `implicitplot3d` using the options `grid=[25,25,25], color=green`. Finally, `display` the quadric with the surface S using the option `orientation=[-10,70]`.

   Notice that the quadric surface is a much better approximation than the tangent plane.

9.7 Lab: Multivariable Max-Min Problems

Objectives: You will learn the *Maple* and `VecCalc` commands used in multivariable max/min and Lagrange multiplier problems.

You are strongly encouraged to work with a partner.

Before Lab: Read chapter 4. Also read the `VecCalc` help page on Multivariable Max-Min Problems which is accessible by executing:

```
> with(VecCalc);
> ?MultiMaxMin
```

Maple Commands: `equate`, `solve`, `fsolve`, `RootOf`, `allvalues`, `eval`, `evalf`, `simplify`, `op`, `plot3d`, `contourplot3d`, `contourplot`, `implicitplot`, `plot`, and `display`.

`VecCalc` Commands: `MF` or `&->` (Make Function), `EF` or `&@` (Evaluate Function), `Grad` (gradient) `Hess` (hessian), and `LPMD` (leading principal minor determinants).

Lab Requirements: Start the `VecCalc` package by executing:

```
> with(VecCalc):  VCalias:
```

Consider one of the following functions. (Your instructor will individually tell you which to use.)

$$f(x, y) = (4x^2 + y^2 - 4)^2 - 8xy + 4x \qquad \text{or} \qquad f(x, y) = 6(y^2 - x^2 - 2xy + 2y)e^{-x^2-y^2/4}$$

Answer the following questions. Number each problem either in a text region or using a *Maple* comment. Where appropriate, you must explain your reasoning in text regions. Save your file now and after each problem. Print your worksheet to turn in as a lab report.

1. **Unconstrained Max-Min Problems:** Find all critical points of f and classify each as a local maxima, a local minima or a saddle point. Here are the suggested steps you should use:

   (a) Define the function$f$ using `MF`.

   (b) Plot the function f to gain a qualitative understanding of the local maxima, local minima and saddle points. Use `contourplot3d` with `orientation=[-90,0]` and `plot3d` with `view=-15..20`.

   (c) Compute the gradient of f and set it equal to the zero vector using `equate`.

   (d) Solve the resulting system of equations to obtain critical points. If necessary, use `allvalues` and `evalf` to resolve any `RootOfs`. Or use `fsolve` with ranges. CHECK your critical points to verify they satisfy the equation $\vec{\nabla} f = \vec{0}$.

   (e) Compute the Hessian of f.

   (f) Analyze each critical point using `LPMD` to determine if it is a local maxima, a local minima or a saddle point. CHECK your classification of each critical point agrees with your qualitative understanding from your plots. If necessary replot the function in the neighborhood of each critical point.

2. **Constrained Max-Min Problems:** Find the absolute maximum and absolute minimum of your function f first on the ellipse $4x^2 + y^2 = 4$ and then inside or on the ellipse. Here are the suggested steps you should use:

   (a) Define the function f and the constraint function $g = 4x^2 + y^2$ using `MF`.

   (b) Use `display` to simultaneously show a `contourplot` of the function f and an `implicitplot` of the constraint equation $g = 4$ to gain a qualitative understanding of the maxima and minima on the constraint. You may want to add additional contours to the `contourplot` by using the `contours=[ ]` option.

   (c) Compute the gradient of f and the gradient of g and construct the Lagrange equations $\vec{\nabla} f = \lambda \vec{\nabla} g$. Also define the constraint equation $g = 4$.

   (d) Solve the resulting system of equations to obtain critical points. If necessary, use `allvalues` and `evalf` to resolve any `RootOfs`. Or use `fsolve` with ranges. CHECK your critical points to verify they satisfy the equations $\vec{\nabla} f = \lambda \vec{\nabla} g$ and $g = 4$.

   (e) Compute the value of f at all critical points on the ellipse to find the absolute maxima and absolute minima on the constraint. CHECK your extrema agree with your qualitative understanding from your plot.

   (f) Plot the critical points found in problem 1 and add them to the implicit plot of the constraint to determine which are inside the ellipse. Then compare all the values to find the absolute maxima and absolute minima inside or on the constraint.

9.8 Lab: A Volume of Desserts

Objectives: You will learn to use *Maple* to compute integrals in two and three dimensional space, in polar, cylindrical and spherical coordinates.

You are strongly encouraged to work with a partner.

Before Lab: Read sections 5.1 and 5.2.

Maple Commands: `Int`, `value`, `simplify`, `plot` and `plot3d`.

`VecCalc` **Commands:** `MF` (Make Function), `Muint` (Display Multiple Integral), `muint/step` (Compute Multiple Integral, Stepwise), `p2r` (polar to rectangular), `r2p` (rectangular to polar), `c2r` (cylindrical to rectangular), `r2c` (rectangular to cylindrical), `s2r` (spherical to rectangular) and `r2s` (rectangular to spherical).

Lab Requirements: Start the `VecCalc` package by executing:

```
> with(VecCalc): VCalias:
```

Answer the following questions. Number each problem either in a text region or using a *Maple* comment. Where appropriate, you must explain your reasoning in text regions. Save your file now and after each problem. Print your worksheet to turn in as a lab report.

1. Consider the chocolate kiss given in cylindrical coordinates by $0 \leq r \leq 1 - \frac{3z^2}{4} + \frac{z^3}{4}$ for $0 \leq z \leq 2$ and $0 \leq \theta \leq 2\pi$.

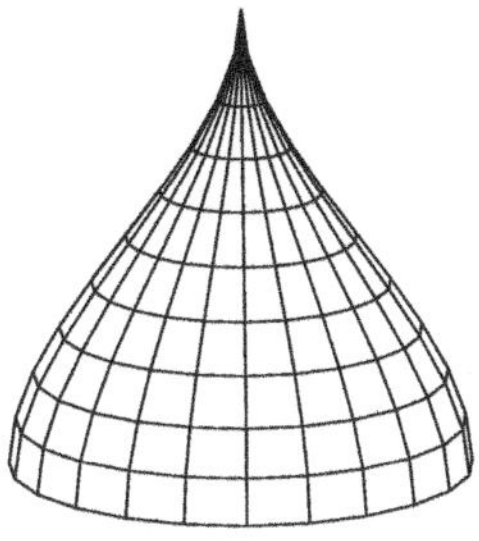

   (a) Find the volume of the chocolate kiss.

   (b) Find the z coordinate of the centroid of the chocolate kiss.
   HINT: Do the integrals in cylindrical coordinates.

2. The top of a pie wedge is given in cylindrical coordinates by $0 \le r \le 4$, $0 \le \theta \le \frac{\pi}{6}$ and $z = 0$. The depth of the pie is given by $z = -1 + \frac{r^2}{16}$

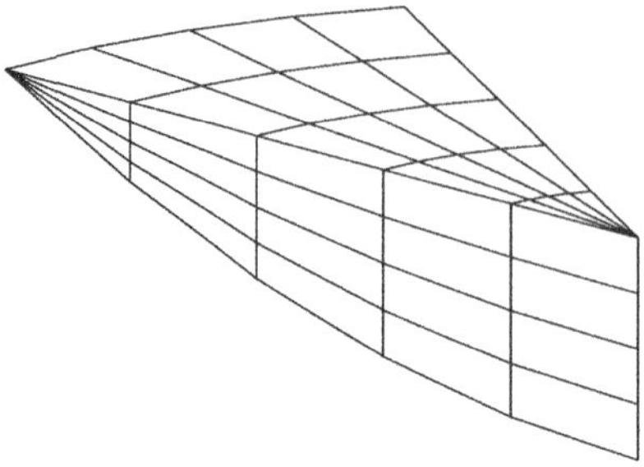

   (a) Find the area of the top of the pie wedge.
   HINT: Use polar coordinates.

   (b) Find the x and y coordinates of the centroid of the top of the pie wedge. Then express the centroid in polar coordinates.
   HINT: Set up the integrals to find the x and y coordinates of the centroid, NOT the r and θ components. Then work in polar coordinates to do the integrals. Finally convert the x and y coordinates of the centroid to polar coordinates.

   (c) Find the volume of the solid pie wedge.
   HINT: Use cylindrical coordinates.

   (d) Find the x, y and z coordinates of the centroid of the solid pie wedge. Then convert to cylindrical coordinates.
   HINT: You need to do 3 dimensional integrals, especially for the z component of the centroid.

3. Consider the orange wedge given in spherical coordinates by $0 \le \rho \le 5$, $0 \le \theta \le \frac{\pi}{6}$ and $0 \le \phi \le \pi$.

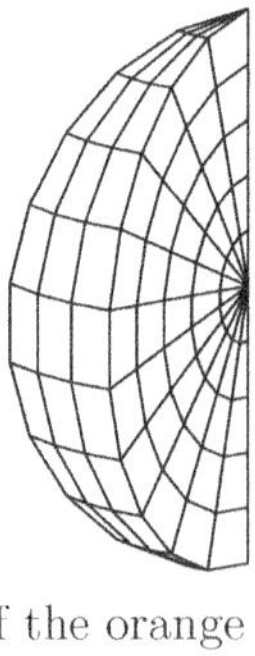

   (a) Find the volume of the orange wedge.

   (b) Find the x, y and z coordinates of the centroid of the orange wedge. Then convert to spherical coordinates.
   HINT: Work in spherical coordinates for doing the integrals while finding the x, y and z components of the centroid.

4. The surface of an apple is given in spherical coordinates by $\rho = 1 - \cos(\phi)$ for $0 \le \theta \le 2\pi$ and $0 \le \phi \le \pi$.

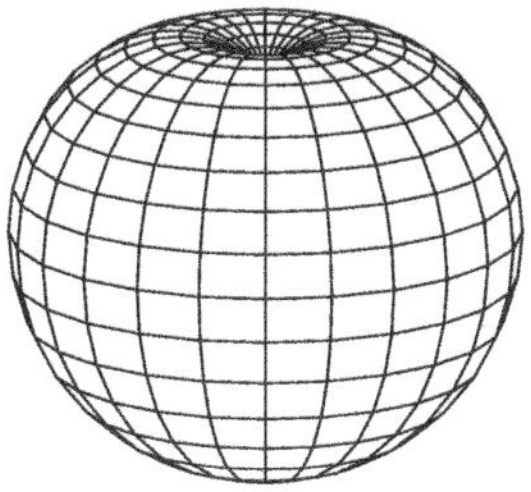

   (a) Find the volume of the apple.

   (b) Find the z coordinate of the centroid of the apple. Then convert to spherical coordinates. HINT: Work in spherical coordinates for doing the integrals while finding the z component of the centroid.

5. (20% EXTRA CREDIT) Plot the chocolate kiss, the pie wedge, the orange slice and the apple. Be sure to plot all surfaces of each. (See section 1.3.3 for plotting parametric surfaces.)

9.9 Lab: Gauss' Law

Objectives: You will learn to use *Maple* to compute volume integrals and surface integrals of a vector field, to compute the divergence of a vector field and to apply Gauss' Theorem. You will also learn about the basic law of electrostatics: Gauss' Law.

Before Lab: Read chapters 5 and 6 and sections 7.2 and 8.5, especially example 7.3.

Maple Commands: `D`, `op`, `Int`, `value` and `simplify`.

`VecCalc` Commands: `MF` or `&->` (Make Function), `EF` or `&@` (Evaluate Function), `Div` (divergence), `Dot` or `&.` (dot product), and `Muint` (Display Multiple Integral).

Background: Gauss' Law is the basic law of electrostatics and applies when the electric charge density ρ_c (in units of charge/unit volume) is independent of time. (The charge is allowed to move but the net charge at each point must remain the same.) The integral of the charge density $Q = \iiint\limits_V \rho_c \, dV$ over a volume V gives the net charge Q in the volume V. Gauss' Law relates the electric field $\vec{E}$ to either the charge density ρ_c or the net charge Q.

The *differential form of Gauss' Law* states $\vec{\nabla} \cdot \vec{E} = 4\pi\rho_c$.

The *integral form of Gauss' Law* states $\iint\limits_S \vec{E} \cdot d\vec{S} = 4\pi Q$.

Here S is any closed surface and Q is the net charge in the volume V whose boundary is the surface S. The differential and integral forms of Gauss' Law are related by Gauss' Theorem as discussed in question 3 below.

Lab Requirements: Start the `VecCalc` package by executing:

```
> with(VecCalc): VCalias:
```

Answer the following questions. Number each problem either in a text region or using a *Maple* comment. Where appropriate, you must explain your reasoning in text regions. Save your file now and after each problem. Print your worksheet to turn in as a lab report.

Consider the following 6 electric fields inside a sphere of radius a: (c is a constant.)

- $\vec{E}_1 = cr^2\vec{r} = \left(cx(x^2+y^2+z^2), cy(x^2+y^2+z^2), cz(x^2+y^2+z^2)\right)$
- $\vec{E}_2 = cr\vec{r} = \left(cx\sqrt{x^2+y^2+z^2}, cy\sqrt{x^2+y^2+z^2}, cz\sqrt{x^2+y^2+z^2}\right)$
- $\vec{E}_3 = c\vec{r} = (cx, cy, cz)$
- $\vec{E}_4 = \dfrac{c\vec{r}}{r} = \left(\dfrac{cx}{\sqrt{x^2+y^2+z^2}}, \dfrac{cy}{\sqrt{x^2+y^2+z^2}}, \dfrac{cz}{\sqrt{x^2+y^2+z^2}}\right)$
- $\vec{E}_5 = \dfrac{c\vec{r}}{r^2} = \left(\dfrac{cx}{x^2+y^2+z^2}, \dfrac{cy}{x^2+y^2+z^2}, \dfrac{cz}{x^2+y^2+z^2}\right)$
- $\vec{E}_6 = \dfrac{c\vec{r}}{r^3} = \left(\dfrac{cx}{(x^2+y^2+z^2)^{3/2}}, \dfrac{cy}{(x^2+y^2+z^2)^{3/2}}, \dfrac{cz}{(x^2+y^2+z^2)^{3/2}}\right)$

Start by telling *Maple* that the constant c and the radius a are positive:

```
> assume(c>0,a>0);
```

1. For each electric field, compute the charge density ρ_c by using the differential form of Gauss' Law. Then integrate the charge density over the solid sphere $x^2+y^2+z^2 \le a^2$ to obtain the net charge Q inside the sphere. The integral should be done in spherical coordinates.

2. For each electric field, recompute the net charge Q inside the sphere by using the integral form of Gauss' Law. The boundary of the solid sphere is the surface $x^2+y^2+z^2 = a^2$ which may be parametrized by $\vec{R}(\theta,\phi) = (a\sin(\phi)\cos(\theta), a\sin(\phi)\sin(\theta), a\cos(\phi))$.

3. Without using *Maple*, explain how the integral form of Gauss' Law may be derived from the differential form of Gauss' Law and Gauss' Theorem. You may type this in a text region or write it out by hand.

4. For one of the electric fields, the charge density is constant. Which one?

5. For one of the electric fields, the net charge Q came out differently when using the differential and integral forms of Gauss' Law. Which one? Why does this not violate Gauss' Theorem? In this case, the physicists regard the integral form of Gauss' Law as giving the correct answer and interpret the net charge Q as a point charge at the origin. Explain why this interpretation is reasonable by looking at the charge density at points other than the origin.

9.10 Lab: Ampere's Law

Objectives: You will learn to use *Maple* to compute line and surface integrals of a vector field, to compute the curl of a vector field and to apply Stokes' Theorem. You will also learn about the basic law of magnetostatics: Ampere's Law.

Before Lab: Read chapter 6 and sections 7.3 and 8.4, especially examples 6.7 and 7.6.

Maple Commands: `D`, `op`, `Int`, `value` and `simplify`. NOTE: In *Maple*, the letter `I` stands for $\sqrt{-1}$. So you will need to use some other symbol for the current, e.g. `Ia, Ib,` etc.

`VecCalc` **Commands:** `MF` or `&->` (Make Function), `EF` or `&@` (Evaluate Function), `Curl` (curl), `Dot` or `&.` (dot product), `Cross` or `&x` (cross product), and `Muint` (Display Multiple Integral).

Background: Ampere's Law is the basic law of magnetostatics and applies when the electric charge density ρ_c (in units of charge/unit volume) is independent of time. (The charge is allowed to move but the net charge at each point must remain the same.) If the velocity field of the charge is $\vec{V}$ (in units of distance/unit time) then the current density is $\vec{J} = \rho_c \vec{V}$ (in units of charge/unit area/unit time). The integral of the current density $I = \iint\limits_S \vec{J} \cdot \vec{dS}$ over a surface S gives the net current I (in units of charge/unit time) which is passing through that surface, positive in the direction of the normal to S. Ampere's Law relates the magnetic field $\vec{B}$ to either the current density $\vec{J}$ or the current I.

The *differential form of Ampere's Law* states $\vec{\nabla} \times \vec{B} = 4\pi \vec{J}$.

The *integral form of Ampere's Law* states $\oint\limits_C \vec{B} \cdot \vec{ds} = 4\pi I$.

Here C is any closed curve and I is the net current passing through any surface S whose boundary is the curve C. The differential and integral forms of Ampere's Law are related by Stokes' Theorem as discussed in question 3 below.

Lab Requirements: Start the `VecCalc` package by executing:

```
> with(VecCalc): VCalias:
```

Answer the following questions. Number each problem either in a text region or using a *Maple* comment. Where appropriate, you must explain your reasoning in text regions. Save your file now and after each problem. Print your worksheet to turn in as a lab report.

Consider the following 6 magnetic fields: (c is a constant and $\vec{r}^{\perp} = (-y, x, 0)$.)

- $\vec{B}_1 = 2c \left|r^{\perp}\right|^2 \vec{r}^{\perp} = \left(-2cy(x^2+y^2), 2cx(x^2+y^2), 0\right)$
- $\vec{B}_2 = 2c \left|r^{\perp}\right| \vec{r}^{\perp} = \left(-2cy\sqrt{x^2+y^2}, 2cx\sqrt{x^2+y^2}, 0\right)$
- $\vec{B}_3 = 2c\vec{r}^{\perp} = (-2cy, 2cx, 0)$
- $\vec{B}_4 = \dfrac{2c\vec{r}^{\perp}}{|r^{\perp}|} = \left(\dfrac{-2cy}{\sqrt{x^2+y^2}}, \dfrac{2cx}{\sqrt{x^2+y^2}}, 0\right)$
- $\vec{B}_5 = \dfrac{2c\vec{r}^{\perp}}{|r^{\perp}|^2} = \left(\dfrac{-2cy}{x^2+y^2}, \dfrac{2cx}{x^2+y^2}, 0\right)$
- $\vec{B}_6 = \dfrac{2c\vec{r}^{\perp}}{|r^{\perp}|^3} = \left(\dfrac{-2cy}{(x^2+y^2)^{3/2}}, \dfrac{2cx}{(x^2+y^2)^{3/2}}, 0\right)$

You wil be integrating these fields over a circle of radius a. So start by telling *Maple* that the constant c and the radius a are positive:

```
> assume(c>0,a>0);
```

1. For each magnetic field, compute the current density $\vec{J}$ by using the differential form of Ampere's Law. Then integrate the current density over the disk $x^2 + y^2 \leq a^2$ in the xy-plane to obtain the net current I passing through the disk. The disk may be parametrized in polar coordinates by $\vec{R}(r, \theta) = (r\cos(\theta), r\sin(\theta), 0)$.

2. For each magnetic field, recompute the net current I passing through the disk by using the integral form of Ampere's Law. The boundary of the disk is the circle $x^2 + y^2 = a^2$ which may be parametrized by $\vec{r}(\theta) = (a\cos(\theta), a\sin(\theta), 0)$.

3. Without using *Maple*, explain how the integral form of Ampere's Law may be derived from the differential form of Ampere's Law and Stokes' Theorem. You may type this in a text region or write it out by hand.

4. For one of the magnetic fields, the current density is constant. Which one?

5. For one of the magnetic fields, the net current I came out differently when using the differential and integral forms of Ampere's Law. Which one? Why does this not violate Stokes' Theorem? In this case, the physicists regard the integral form of Ampere's Law as giving the correct answer and interpret the net current I as a current moving along the z-axis. Explain why this interpretation is reasonable by looking at the current density at points not on the z-axis.

9.11 Lab: Interpretation of the Divergence

Objectives: You will learn an integral formula for the divergence of a vector field which provides an interpretation of the divergence. Then you will prove it using Gauss' Theorem and use it to compute several divergences.

Before Lab: Read subsection 6.2.6 and sections 7.2 and 8.5.

Maple Commands: `fieldplot3d`, `D`, `op`, `Limit` and `value`.

`VecCalc` **Commands:** `MF` or `&->` (Make Function), `EF` or `&@` (Evaluate Function), `Div` (divergence), `Dot` or `&.` (dot product), `Cross` or `&x` (cross product), and `Muint` (Display Multiple Integral).

Background: Given a 3-dimensional vector field $\vec{F} = (F_1, F_2, F_3)$, the following integral formula for the divergence of $\vec{F}$ gives the value of the function div$\vec{F}$ at a point P. To do this, let $S_\rho(P)$ be the sphere centered at P of radius ρ oriented with the outward normal. Then the value of div$\vec{F}$ at P is

$$(\text{div}\vec{F})(P) = \lim_{\rho \to 0} \left[\frac{3}{4\pi\rho^3} \iint\limits_{S_\rho(P)} \vec{F} \cdot d\vec{S} \right]$$

In this formula, the integral computes the outward expansion of the vector field $\vec{F}$ through the sphere $S_\rho(P)$. The integral is then divided by the volume of the sphere $\frac{4}{3}\pi\rho^3$. The limit thus computes the expansion per unit volume for smaller and smaller spheres. Thus, if $\vec{F}$ is interpreted as the velocity of a fluid, then the expansion integral measures the amount of that fluid flowing out of the sphere. So (div$\vec{F}$)(P) may be interpreted as the amount of that fluid which is "coming out" of the point P. Computationally, the sphere centered at $P = (a, b, c)$ of radius ρ may be parametrized by

$$\vec{R}(\theta, \phi) = (a + \rho \sin\phi \cos\theta, b + \rho \sin\phi \sin\theta, c + \rho \cos\phi)$$

Be sure to check the direction of the normal and reverse it if necessary.

Lab Requirements: Start the `VecCalc` package by executing:

```
> with(VecCalc):  VCalias:
```

Answer the following questions. Number each problem either in a text region or using a *Maple* comment. Where appropriate, you must explain your reasoning in text regions. Save your file now and after each problem. Print your worksheet to turn in as a lab report.

1. Justify the integral formula for the divergence by using Gauss' Theorem. You may assume that div$\vec{F}$ is continuous, so that its value inside a small sphere may be approximated by its value at the center of the sphere. You should not use *Maple* to do this.

2. Plot each of the following vector fields for $-5 \le x \le 5$, $-5 \le y \le 5$, $-5 \le z \le 5$. Then compute the divergence at the point $P = (a, b, c)$ using both the derivative formula (using `Div`) and the integral formula. Check that the answers are the same.

   (a) $\vec{F} = (x^3, y^3, z^3)$

   (b) $\vec{G} = (x, xy, xyz)$

   (c) $\vec{u} = (xy - xz, yz - yx, zx - zy)$

   (d) $\vec{v} = (0, 0, \ln(z))$

   (e) $\vec{w} = (x^2yz, xy^2z, xyz^2)$

9.12 Lab: Interpretation of the Curl

Objectives: You will learn an integral formula for the curl of a vector field which provides an interpretation of the curl. Then you will prove it using Stokes' Theorem and use it to compute several curls.

Before Lab: Read subsection 6.1.4 and sections 7.3 and 8.4.

Maple Commands: `fieldplot3d`, `D`, `Limit`, `Int` and `value`.

`VecCalc` **Commands:** `MF` or `&->` (Make Function), `EF` or `&@` (Evaluate Function), `Curl` (curl), and `Dot` or `&.` (dot product).

Background: Given a 3-dimensional vector field $\vec{F} = (F_1, F_2, F_3)$, its curl is also a vector field. The following integral formula for the curl of $\vec{F}$ gives the value at a point P of the quantity $\hat{N} \cdot \text{curl}\vec{F}$ for an arbitrary unit vector $\hat{N}$. Then the components of $\text{curl}\vec{F}$ may be found by taking $\hat{N}$ to be successively $\hat{\imath}$, $\hat{\jmath}$ and $\hat{k}$. To give the formula, let $C_{(\rho,\hat{N})}(P)$ be the circle centered at P of radius ρ which lies in the plane through P with normal $\hat{N}$. Then the value of $\hat{N} \cdot \text{curl}\vec{F}$ at P is

$$(\hat{N} \cdot \text{curl}\vec{F})(P) = \lim_{\rho \to 0} \left[\frac{1}{\pi\rho^2} \oint_{C_{(\rho,\hat{N})}(P)} \vec{F} \cdot \vec{ds} \right]$$

where the circle is traversed counterclockwise as seen from the tip of the normal $\hat{N}$.

In this formula, the integral computes the circulation of the vector field $\vec{F}$ around the circle $C_{(\rho,\hat{N})}(P)$. The integral is then divided by the area of the circle $\pi\rho^2$. The limit thus computes the circulation per unit area for smaller and smaller circles. Thus, if $\vec{F}$ is interpreted as the velocity of a fluid, then $(\hat{N} \cdot \text{curl}\vec{F})(P)$ may be interpreted as the rate that the fluid circulates about the line through the point P in the direction $\hat{N}$. If $(\hat{N} \cdot \text{curl}\vec{F})(P) > 0$, then the fluid circulates counterclockwise as seen from the tip of $\hat{N}$. If $(\hat{N} \cdot \text{curl}\vec{F})(P) < 0$, then the fluid circulates clockwise.

Also notice that $(\hat{N}\cdot\text{curl}\vec{F})(P)$ is a maximum when $\hat{N}$ points in the same direction as $(\text{curl}\vec{F})(P)$. Thus $(\text{curl}\vec{F})(P)$ points along the axis of rotation at P and its magnitude $\left|(\text{curl}\vec{F})(P)\right|$ is the rate of rotation at P.

Computationally, if $P = (a, b, c)$, then the three components of $(\text{curl}\vec{F})(P)$ are given by

$$(\text{curl}\vec{F})_1(P) = \lim_{\rho\to 0} \frac{1}{\pi\rho^2} \oint_{C_1} \vec{F}\cdot\vec{ds}$$

$$(\text{curl}\vec{F})_2(P) = \lim_{\rho\to 0} \frac{1}{\pi\rho^2} \oint_{C_2} \vec{F}\cdot\vec{ds}$$

$$(\text{curl}\vec{F})_3(P) = \lim_{\rho\to 0} \frac{1}{\pi\rho^2} \oint_{C_3} \vec{F}\cdot\vec{ds}$$

where the circles C_1, C_2 and C_3 are parametrized by

$$C_1(\theta) = P + \rho\cos(\theta)\hat{\jmath} + \rho\sin(\theta)\hat{k} = (a, b + \rho\cos(\theta), c + \rho\sin(\theta))$$
$$C_2(\theta) = P + \rho\cos(\theta)\hat{k} + \rho\sin(\theta)\hat{\imath} = (a + \rho\sin(\theta), b, c + \rho\cos(\theta))$$
$$C_3(\theta) = P + \rho\cos(\theta)\hat{\imath} + \rho\sin(\theta)\hat{\jmath} = (a + \rho\cos(\theta), b + \rho\sin(\theta), c)$$

Lab Requirements: Start the `VecCalc` package by executing:

```
> with(VecCalc): VCalias:
```

Answer the following questions. Number each problem either in a text region or using a *Maple* comment. Where appropriate, you must explain your reasoning in text regions. Save your file now and after each problem. Print your worksheet to turn in as a lab report.

1. Justify the integral formula for the curl by using Stokes' Theorem. You may assume that $\text{curl}\vec{F}$ is continuous, so that its value inside a small sphere may be approximated by its value at the center of the sphere. You should not use *Maple* to do this.

2. Plot each of the following vector fields for $-5 \le x \le 5$, $-5 \le y \le 5$, $-5 \le z \le 5$. Then compute the curl at the point $P = (a, b, c)$ using both the derivative formula (using `Curl`) and the integral formula. Check that the answers are the same.

   (a) $\vec{F} = (z^2x, x^2y, y^2z)$

   (b) $\vec{G} = (x^4 + y^3 + z^2, y^4 + z^3 + x^2, z^4 + x^3 + y^2)$

   (c) $\vec{v} = (3x^2\sin(y), x^3\cos(y) - z^3\sin(y), 3z^2\cos(y))$

Chapter 10

Projects

This chapter contains a collection of projects on vector calculus. They are divided into two groups. The first group involve Multivariable Differentiation while the second group also involve Multivariable Integration. In a one semester course, we recommend that the students be required to do two such projects, probably one from each group. Normally the students would work in groups of four students and different groups would work on different projects. Each group would have from two to four weeks to complete the project and must turn in an extensive project report. The report should be graded on mathematics, *Maple* and documentation.

Projects on Vectors and Multivariable Differentiation

- 10.1 Totaling Gravitational Forces
- 10.2 Animate a Curve
- 10.3 Newton's Method in 2 Dimensions
- 10.4 Gradient Method of Finding Extrema
- 10.5 Seeing a Blimp
- 10.6 The Trash Dumpster
- 10.7 Generalized Diameter
- 10.8 Locating an Apartment

Projects on Multivariable Integration

- 10.9 The Skimpy Donut
- 10.10 The Volume Between a Surface and its Tangent Plane
- 10.11 The Hypervolume of a Hypersphere
- 10.12 Average Temperatures
- 10.13 The Center of Mass of Planet X
- 10.14 Steradian Measure

10.1 Project: Totaling Gravitational Forces

In physics, Newton's Law of Gravity says that a point mass M is attracted to a point mass m by the force $\vec{F} = \dfrac{GMm}{|\vec{r}|^3}\vec{r}$, where G is Newton's gravitational constant, $\vec{r}$ is the vector from M to m, and $|\vec{r}|$ is the length of $\vec{r}$. More generally, a point mass M is attracted to a collection of point masses $m_1, m_2, \ldots, m_k$ by the force $\vec{F} = \displaystyle\sum_{i=1}^{k} \frac{GMm_i}{|\vec{r}_i|^3}\vec{r}_i$, where $\vec{r}_i$ is the vector from M to m_i.

Suppose a point mass M is located on the y-axis at $P = (0, Y)$ for a positive number Y. Also suppose $2n$ point masses, each with mass m, are located on the x-axis at the points $Q_i = (i\Delta x, 0)$ for a positive number Δx and $i = -n, -n+1, -n+2, \ldots, -1, 1, \ldots, n-2, n-1, n$. (NOTE: There is no mass at the origin.)

1. Draw a picture showing the line segments from P to each of the Q_i's when $Y = 4$, $n = 3$ and $\Delta x = 1$. (See lab 9.1 for an example of plotting dot-to-dot pictures.)
2. Find a series expression for the magnitude of the total gravitational force on M due to the $2n$ other masses with general n and Δx. (NOTE: You can't sum it.)
3. Is the magnitude of the total gravitational force on M finite if n goes to infinity? Use a convergence or divergence test to explain why or why not.
4. With $n = 3$ and $\Delta x = 1$, find the values of Y which maximize and minimize the magnitude of the total gravitational force on M due to the 6 other masses. Check with a plot.

10.2 Project: Animate a Curve

Write a *Maple procedure* (See Appendix C.) which animates the graph of a curve $\vec{r}(t)$ in 3 dimensions. (Read the help page `?plots,display`, especially the option `insequence=true`.) The inputs should be the curve function, the range for the parameter and the number of plots. The output should be the animated plot. At the tip of the curve add one of the following to your animation:

1. The velocity vector $\vec{v}(t)$ and the acceleration vector $\vec{a}(t)$.
2. The unit tangent vector $\hat{T}(t)$, the unit principal normal vector $\hat{N}(t)$ and the unit binormal vector $\hat{B}(t)$.
3. The osculating circle. This is the circle in the plane of $\vec{v}(t)$ and $\vec{a}(t)$ which best fits the curve at $\vec{r}(t)$. Thus, its center is in the direction $\hat{N}(t)$ from $\vec{r}(t)$ and its radius is $\dfrac{1}{\kappa(t)}$ where $\kappa(t)$ is the curvature.

Use your procedure to animate a few curves using about 15 plots. Document your procedure to explain how it works and write a help page to explain how it may be used.

10.3 Project: Newton's Method in 2 Dimensions

The ordinary Newton's Method uses the linear approximation to find an approximate solution to an equation of the form $f(x) = 0$. Basically, if x_0 is an initial approximation to the solution, then the tangent line to $y = f(x)$ at $x = x_0$ intersects the x-axis at a point $(x_1, 0)$ and x_1 is usually a better approximation to the solution than x_0. So the process can be iterated using x_1 as the new initial approximation. A short derivation shows that at each stage

$$x_{i+1} = x_i - \frac{f(x_i)}{f'(x_i)}$$

This may be automated in *Maple* by defining the function

```
> newt:= x -> evalf(x - f(x)/Df(x));
```

This assumes that f and its derivative Df have been defined in arrow notation. Further, it is often useful to plot $y = f(x)$ to get an initial approximation to the solution and to set `Digits` to one more than the desired number of digits accuracy.

Example: 10.1. Solve the equation $\cos(x) = x$ to 15 digits of accuracy.

SOLUTION: We set the digits, define the function and compute its derivative:

```
> Digits:=16;
> f:=x -> cos(x)-x;
> Df:=D(f);
```

To get an initial approximation, we plot the function:

```
> plot(f(x),x=-Pi..Pi);
```

and observe the initial approximation should be $x = .8$. We can now use 5 iterations of Newton's method to get the solution:

```
> newt(.8); newt(%); newt(%); newt(%); newt(%);
```

We now turn to the 2-dimensional Newton's Method. This uses the linear approximation to find an approximate solution to a pair of equations of the form $f(x, y) = 0$ and $g(x, y) = 0$. Basically, if (x_0, y_0) is an initial approximation to the solution, then the tangent plane to $z = f(x, y)$ at $(x, y) = (x_0, y_0)$ and the tangent plane to $z = g(x, y)$ at $(x, y) = (x_0, y_0)$ intersect the xy-plane at a common point $(x_1, y_1, 0)$ and (x_1, y_1) is usually a better approximation to the solution than (x_0, y_0). So the process can be iterated using (x_1, y_1) as the new initial approximation. A short derivation shows that at each stage

$$x_{i+1} = x_i - \frac{f g_y - f_y g}{f_x g_y - f_y g_x} \qquad \text{and} \qquad y_{i+1} = y_i - \frac{f_x g - f g_x}{f_x g_y - f_y g_x} \tag{*}$$

where the functions f and g and their partial derivatives f_x, f_y, g_x and g_y are all evaluated at (x_i, y_i).

1. Derive the equations (*). You should use *Maple* to construct the tangent planes and to solve for the intersection of these planes with the xy-plane. Your derivation should read like a page from a calculus text.

2. Construct a single *Maple* function called `newt2d` which acts on an initial approximation (x, y) and produces the next approximation.

3. (Optional) To improve your project, write a *Maple procedure* (See Appendix C.) which will automatically control the iterations of `newt2d`. The procedure should take as arguments, the functions f and g, the initial point, the number of digits of accuracy desired and the maximum number of iterations to allow (to prevent an infinite loop).

4. Use your *Maple* function `newt2d` or your *Maple* procedure to find all solutions to each of the following pairs of equations. You will need to plot the two equations using `implicitplot` to get an initial approximation to each solution. Give your answers to 25 digits of accuracy. (See `?Digits`.) You can use `fsolve` to check you solutions.

   (a) $x + y - \cos(x) + \sin(y - 1) = 0$ and $x^4 + y^4 - 2xy = 0$

   (b) $5x - 3y = -2$ and $2x - 2y = -3$

   (c) $y + (x - 2)^2 = 25$ and $x^4 - y^4 = 16$

   (d) $y^3x - x^3y + x^2y^2 = 7$ and $2x^4 + 3y^4 = 74$

10.4 Project: Gradient Method of Finding Extrema

Write a *Maple procedure* (See Appendix C.) which finds an approximation to a local maximum or local minimum of a given function using the gradient method (also called Cauchy's method or the method of steepest ascent or descent). Use for several functions.

Cauchy's Gradient Method or the Method of Steepest Ascent Suppose you want to find a local maximum of a function $f(\vec{X})$ near an initial point $\vec{P}_0$. Recall the gradient vector points in the local direction of maximum increase of the function but it may not point directly at the top of the hill. So, if you move from $\vec{P}_0$ in the direction of the gradient of f at $\vec{P}_0$, then the function will increase, at least initially. Hence, construct the line $\vec{X}(t) = \vec{P}_0 + t\,\vec{\nabla} f(\vec{P}_0)$ and restrict the function to this line by forming the composition. $g(t) = f(X(t)) = f(\vec{P}_0 + t\,\vec{\nabla} f(\vec{P}_0))$ Let $\vec{P}_1$ be the first maximum of $g(t)$ for $t > 0$. Now $\vec{P}_1$ is a local maximum of g but may not be a local maximum of f because $\vec{\nabla} f(\vec{P}_0)$ only points *locally* uphill. However, $f(\vec{P}_1)$ is bigger than $f(\vec{P}_0)$. So iterate this process with $\vec{P}_1$ as the new initial point (hopefully) getting closer and closer to the local maximum.

- The inputs to your procedure should be the function, the initial guess, the maximum number of iterations (to prevent an infinite loop), the desired tolerance and a parameter to say whether the program should look for a maximum or a minimum. The output should be the coordinates of the extremum which may be plugged into the function to obtain the extreme value.
- Document your procedure to explain how it works. Write a help page to explain how it may be used.
- Before you turn the algorithm into a procedure or even automate it using a `for/while/do` loop, be sure your algorithm and your *Maple* code work step by step on one or two of the sample functions.

- To find a maximum you move in the direction of the gradient: $\vec{\nabla} f$. To find a minimum you move in the direction of the negative of the gradient: $-\vec{\nabla} f$.
- Along the line in the direction of the gradient, there may be several critical points. Which one do you want? To isolate this critical point, use an interval in the `fsolve` command and/or use the `max` or `min` commands to find the largest or smallest of a list of numbers.
- Your input function may be an expression or an arrow defined function and your max/min parameter may be a number or string, but explain which in the documentation.
- Your tolerance may measure the distance moved between two successive iterations or the change in the value of the function between two successive iterations or both. Your documentation must explain this.
- Read the help pages on `?proc`, `?options`, `?for` and `?if`. To debug a procedure, add `option trace;` at the beginning of the procedure and add `print` statements.
- (Optional) To improve your project, generalize the procedure to functions of 3 or more variables.

Use your procedure to find *all* local extrema of each of the following functions to 10 digits of accuracy. To help find each initial guess, plot the function with `plot3d` and `contourplot3d`. Check out the `color=f` and `contours` options. Compare your results to those given by `fsolve`.

1. $f(x, y) = 5 - (x - 3)^2 - 2(y - 4)^2$ (Start from the point $(2, 3)$.)
2. $f(x, y) = ((x - 1)^4 + (y - 2)^4 - 4)^2 + 5xy - 12x - 2y$ (There are 3 extrema.)
3. $f(x, y) = x^2 - 2x + y^3 - y + 10\sin(x)\cos(y)$ (There are 4 extrema.)

10.5 Project: Seeing a Blimp

A blimp[1] has the shape and location given by the ellipsoid

$$\left(\frac{x-250}{28}\right)^2+\left(\frac{y-750}{96}\right)^2+\left(\frac{z-1200}{30}\right)^2=1.$$

What is the angular size of the blimp as seen from your eye which is located at the origin, O. In other words, find two points, P_1 and P_2, on the blimp, so that the angle between the vectors $\overrightarrow{OP_1}$ and $\overrightarrow{OP_2}$ is as large as possible. Plot the ellipsoid and the vectors $\overrightarrow{OP_1}$ and $\overrightarrow{OP_2}$. Orient the plot so you can see the angle between $\overrightarrow{OP_1}$ and $\overrightarrow{OP_2}$.

HINTS:

- Solve the problem by one (or more) of the three methods discussed in section 4.2: Lagrange Multipliers, Parametrizing the Constraints and Eliminating Variables.
- When needed, the ellipsoid may be parametrized by

$$\vec{R}(\phi,\theta)=(250+28\sin(\phi)\cos(\theta),750+96\sin(\phi)\sin(\theta),1200+30\cos(\phi)).$$

- Since the ellipse is longest in the y-direction, you would expect the points P_1 and P_2 to be close to the endpoints of the y-diameter; i.e. close to the points $P_{1o}=\langle 250,654,1200\rangle$ and $P_{2o}=\langle 250,846,1200\rangle$. Find the angle between the vectors $\overrightarrow{OP_{1o}}$ and $\overrightarrow{OP_{2o}}$. The actual angle should be slightly larger than this.
- If you do nothing to restrict the `fsolve` command, *Maple* is most likely to miss the maximum and find a minimum of 0 angle which occurs whenever $P_1=P_2$. So in the `fsolve` command, restrict the ranges of the variables so that P_1 is close to P_{1o} and P_2 is close to P_{2o}.
- If you eliminate variables, express y_1 in terms of x_1 and z_1 in the half of the ellipse near P_{1o} and express y_2 in terms of x_2 and z_2 in the half near P_{2o}.
- In the Lagrange method, eliminate the Lagrange multipliers before using `fsolve`.
- In the parametric method, use expressions and `diff` instead of arrow-defined functions and `Grad`.

(Optional) To improve your project, estimate the size of the lettering on the side of the blimp to be visible to someone at the origin who has 20-20 vision. To do this you will need to calculate the angular size of the 20-20 line on an eye chart as seen from 20 feet away.

[1] The dimensions and altitude of the Goodyear blimp were taken from http://www.goodyearblimp.com/ on the FAQ tab under "ABOUT FLYING" and "BLIMP CONSTRUCTION".

10.6 Project: The Trash Dumpster

You are the mathematics consultant for a company which makes trash dumpsters, the big kind you see outside a dorm or apartment complex. Go outside and find one which is not just a rectangular solid and has some type of hinged lid covering part of the top. If you cannot find one with a lid, pretend that a lid covers the front portion of the top for the full width of the dumpster. This is the kind of trash dumpster your company currently manufactures (hereafter called the original dumpster). Draw a diagram of the original dumpster and take its measurements. Note which edges are folded and which are welded.

Your boss has asked you to redesign the dumpster to minimize the cost, but with the following constraints: (You may modify any of these restrictions to fit your geometry, but you must explain in your report.)

- You must maintain the basic shape of the dumpster but you may change the lengths.
- You must maintain the volume of the dumpster to hold the same amount of trash.
- You must maintain the area of the lid so the dumpster may be emptied in the same manner.
- You may need to restrict the ratio of some lengths to prevent the geometry from changing. You should only do this if the minimization process causes some length to go to zero, thereby changing the geometry. You must document this in your report.
- The base is made of 10 gauge steel sheet metal (.1345 in thick) which costs \$1.99 per ft^2.
- The sides, top and lid are made of 12 gauge steel sheet metal (.1046 in thick) which costs \$1.49 per ft^2.
- Welding costs \$0.19 per ft.
- The hinge for the lid costs \$0.29 per foot.
- Cutting and folding the sheet metal are fixed costs which are independent of length. So they do not need to be included in the cost.

You need to write a report presenting your suggestions with a cover letter for the company president and the technical appendix for the engineers. You should include the original cost and dimensions, the final cost and dimensions and the percent savings in the cost. To organize your work, you should follow the following steps:

1. Draw a diagram of the dumpster. Describe it and pick variable names for each of the lengths.
2. Write formulas for the general cost and volume of the dumpster and the area of the lid.
3. Plug in your measurements to find the original cost, volume and area of the lid.
4. Write out the constraints on the volume and area.
5. Minimize the cost. In your `fsolve` command include ranges to guarantee the variables are positive and the `avoid` option to look for multiple solutions.
6. If some length goes to zero, go back to step 4 and add a constraint on the ratio of that length to some other length.
7. Discuss your results.

10.7 Project: Generalized Diameters

You already know that the diameter of a circle or a sphere is the length of any secant line through the center. More generally:

DEFINITION: The diameter of a closed bounded object in n-dimensional Euclidean space is defined to be the largest distance between any two points in the object.

Plot each of the following objects. Find the diameter by maximizing the square of the distance between two points on the object. Then add the diameter line segment to your plot.

1. 2D: An ellipse: $x^2 + 3y^2 + xy = 16$
   Plot using **implicitplot** with **scaling=constrained**. Use Lagrange multipliers with two constraints. (See subsection 4.2.4.)

2. 2D: A quartic curve: $x^4 + y^4 + x + 4y = 16$
   Plot using **implicitplot** with **scaling=constrained**. Use Lagrange multipliers with two constraints. There are 3 pairs of points which are obvious candidates for the ends of the diameter. Find all 3 pairs and evaluate the distances to see which is largest.

3. 2D: A three petal rose parametrized as: $x = \cos(3t)\cos(t)$ and $y = \cos(3t)\sin(t)$
   Plot using a parametric **plot** with **scaling=constrained**. Express the square of the distance in terms of the parameters.

4. 3D: An ellipsoid: $x^2 + 2y^2 + 3z^2 + xy + 2xz + 3yz = 16$
   Plot the surface using **implicitplot3d** with **scaling=constrained, axes=framed, style=wireframe**. Plot the diameter using **spacecurve** with **thickness=5**. Use Lagrange multipliers with two constraints.

5. 3D: A deformed sphere parametrized as: $x = \big(1 + .3\cos(3\phi)\sin(\theta)\big)\sin(\phi)\cos(\theta)$
   $y = \big(1 + .3\cos(3\phi)\sin(\theta)\big)\sin(\phi)\sin(\theta)$ $z = \big(1 + .3\cos(3\phi)\ sin(\theta)\big)\cos(\phi)$
   Plot the surface using a parametric **plot3d** with **scaling=constrained, axes=framed, style=wireframe**. Plot the diameter using **spacecurve** with **thickness=5**. Express the square of the distance in terms of the parameters. There are 2 candidates for the ends of the diameter. Find both and evaluate the distances to see which is largest.

HINTS:

- Since both points satisfy the same constraint or parametrization, *Maple* is most likely to miss the maximum and find a minimum with $P_1 = P_2$. So in the **fsolve** command, restrict the ranges of the variables so P_1 and P_2 are near the endpoints of the diameter line segment.
- Remember that the distance is the square root of the function you maximized.

10.8 Project: Locating an Apartment

Upon moving to a new city, you want to find an apartment which is conveniently located relative to your school, your place of work and the shopping mall. These are located at

$$S = \langle -2, -2 \rangle \qquad W = \langle 4, 1 \rangle \qquad M = \langle 1, 5 \rangle$$

respectively. If your apartment is at $A = \langle x, y \rangle$ find the location of your apartment which minimizes $f = |\vec{AS}| + |\vec{AW}| + |\vec{AM}|$. Here $|\vec{AS}|$ is the distance from your apartment to school (i.e. the length of the vector $\vec{AS}$) and similarly for $|\vec{AW}|$ and $|\vec{AM}|$.

In the course of solving this problem, you should answer the following questions:

1. Compute the gradient of $|\vec{AS}|$ and express your answer in terms of the vector $\vec{AS}$. In particular, how are their directions related, how are their magnitudes related?
2. Draw a contour plot of $|\vec{AS}|$. What does it say about the direction and magnitude of the gradient $|\vec{AS}|$.
3. Find the point A which minimizes f.
4. Plot the three vectors $\vec{AS}$, $\vec{AW}$ and $\vec{AM}$ together with the triangle $\triangle SWM$, using the `plot` option `scaling=constrained`. (See lab 9.1 for an example of plotting dot-to-dot pictures.)
5. Give a geometric condition on the three vectors $\vec{AS}$, $\vec{AW}$ and $\vec{AM}$ which characterizes the point A which minimizes f which you feel would be true even if S, W or M were moved. Verify it for the given values.

- Do either #6 or #7 or both:

6. What happens if the point M is moved so that the angle $\angle SWM$ is greater than $135°$? Move M until you find the critical angle. Animate your plots.
7. Prove the geometric condition you found in #5 for general position of S, W and M. It may be useful to use your results from #1.

10.9 Project: The Skimpy Donut

You are the mathematics consultant for a donut company which makes donuts which have a thin layer of chocolate icing covering the entire donut. One day you decide to point out that the company might cut costs on chocolate icing if they keep the volume (and hence weight) of the donut fixed but adjust the shape of the donut to minimize the surface area. Alternatively, they could advertise extra icing by maximizing the surface area. You need to write a report presenting your ideas which can be read by both the company president and the technical engineers.

A donut has the shape of a torus which is specified by giving a big radius a from the center of the hole to the center of the ring and a small radius b which is the radius of the ring, as shown in the figure. A typical donut might have $a = 5$ cm and $b = 3$ cm.

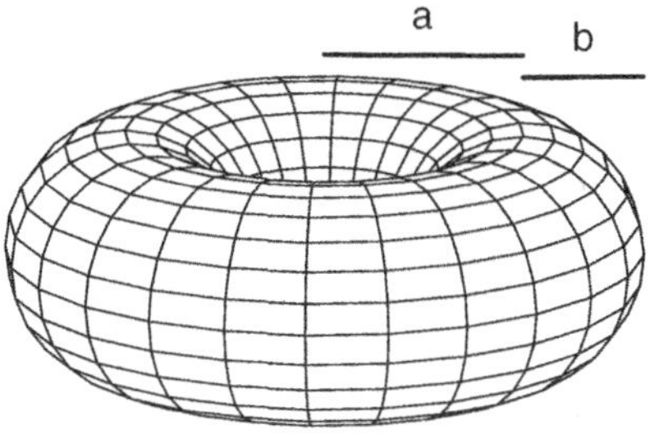

Your job is to determine the values of a and b which extremize the surface area while keeping the volume fixed at the volume of the typical donut mentioned above.

1. The surface of a torus satisfies the equation
$$(r-a)^2 + z^2 = b^2$$
in cylindrical coordinates where, of course, $b \leq a$.
   (a) Compute the volume V of the torus as a function of a and b.
   HINT: `assume(a>b,b>0)` Then, integrate in cylindrical coordinates.
   (b) Check that the volume of the typical donut with $a = 5$ cm and $b = 3$ cm is $V = 90\pi^2$ cm^3 ≈ 888 cm^3.

2. The surface of the torus can also be parametrized as
$$\begin{aligned} x &= (a + b\cos\phi)\cos\theta \\ y &= (a + b\cos\phi)\sin\theta \\ z &= b\sin\phi \end{aligned} \qquad \text{for} \qquad \begin{array}{c} 0 \leq \phi \leq 2\pi \\ 0 \leq \theta \leq 2\pi \end{array}$$
Here, θ represents the angle around the circle of radius a and ϕ represents the angle around the circle of radius b.
   (a) Plot the donut using a 3 dimensional parametric plot.
   (b) Compute the surface area S of the torus as a function of a and b.
   HINT: Do a surface integral in θ and ϕ.
   (c) Check that the surface area of the typical donut with $a = 5$ cm and $b = 3$ cm is $S = 60\pi^2$ cm^2 ≈ 592 cm^2.

3. Keep the volume fixed at $V = 90\pi^2$ cm^3 and find the values of a, b and S which minimize and maximize the surface area S. (Apply the second derivative test to any critical point in the interior and check the values at the endpoints.)

4. Write a letter to the CEO of the donut company summarizing your results (including minimum and maximum dimensions, minimum and maximum plots and the percent savings or extra cost). Anything you say in this report must be documented in an appendix of Maple computations for the engineers.

10.10 Project: The Volume Between a Surface and Its Tangent Plane

In this project, you will be finding the tangent plane to a surface for which the volume between the surface and the tangent plane is a minimum.

1. Consider the surface
$$z = f(x, y) = 4x^2 + y^4 + 2x^2y^2.$$
Verify that the surface is everywhere concave up on the square $0 < x < 1, \ 0 < y < 1$.

NOTE: A function $f(x, y)$ is everywhere concave up if $D = f_{xx}f_{yy} - {f_{xy}}^2 > 0$ and $f_{xx} > 0$ everywhere. It is everywhere concave down if $D = f_{xx}f_{yy} - {f_{xy}}^2 > 0$ and $f_{xx} < 0$ everywhere.

2. Find its tangent plane at a general point $(a, b, f(a, b))$. Plot the function and its tangent plane at $(a, b) = (.3, .3)$
3. Compute the volume between the surface and its general tangent plane above the region R which is the square $0 \le x \le 1, \ 0 \le y \le 1$. Call this volume $V(a, b)$.
4. Find the point (a, b) for which the volume $V(a, b)$ is a minimum. Be sure to apply the second derivative test to verify that your critical point is a minimum.
5. Repeat steps 1-4 for two or three other functions $f(x, y)$. Use interesting functions, not just polynomials with at least one concave up and one concave down. Check the concavity. While computing the volume, you may need to use `expand(%)` before `value(%)`.
6. What do you conjecture?
7. Prove your conjecture by repeating steps 1-4 for an undefined function f. Before solving for (a, b) you will need to give names to the partial derivatives of f using `subs`.
8. What happens to your conjecture if you change the region R? Try some shapes other than a rectangle or a circle!

10.11 Project: The Hypervolume of a Hypersphere

In this project, you will determine the hypervolume enclosed by a hypersphere in $\mathbb{R}^n$ whose equation is: $x_1^2 + x_2^2 + \cdots + x_n^2 = R^2$.

1. Draw the circle $x^2 + y^2 = 1$, using a parametric `plot` or an `implicitplot` with `scaling=constrained`. Compute the area enclosed by the circle $x^2 + y^2 = R^2$ using a double integral in polar coordinates. Repeat using a double integral in rectangular coordinates with x as the inner integral and y as the outer integral. Write your answer as an arrow defined function $V_2(R)$ where V_2 means 2-dimensional volume or area. (In Maple, you enter V_2 as `V[2]`.)

2. Draw the sphere $x^2+y^2+z^2=1$, using a parametric `plot3d` or an `implicitplot3d` with `scaling=constrained`. (You may wish to experiment with various 3-D plotting options.) Compute the volume enclosed by the sphere $x^2+y^2+z^2=R^2$ using a triple integral in spherical coordinates. Repeat using a triple integral in rectangular coordinates with z as the outer integral. Write your answer as an arrow defined function $V_3(R)$ where V_3 means 3-dimensional volume.

- We now leave the earthly realm and journey into n-dimensional space with $n > 3$. Being a three-dimensional being, you cannot visualize objects in these higher dimensional spaces. Like a pilot passing his final flight test, you must rely on your wits and your instruments -- in this case Maple. Take a food break before taking this next step. ... Where did you go and what did you eat?

3. Compute the 4-dimensional hypervolume enclosed by the hypersphere $x^2+y^2+z^2+w^2=R^2$ using a quadruple integral in rectangular coordinates in $\mathbb{R}^4$. Write your answer as an arrow defined function $V_4(R)$ where V_4 means 4-dimensional volume.

4. For $n=5,6,\ldots,10$, find the n-dimensional hypervolume of the n-dimensional hypersphere $\sum_{k=1}^{n} x_k^2 = R^2$ in $\mathbb{R}^n$. Write your answer as an arrow defined function $V_n(R)$ where V_n means n-dimensional volume.

   HINT: After doing the case for $n=5$, you may get *very* tired of typing in all those limits of integration! As an alternative, notice that the 3-dimensional ball of radius R may be sliced into thin disks perpendicular to the z-axis with varying radii r. Computationally, the triple integral for $V_3(R)$ may be written as a single integral over z of $V_2(r)$ with r varying as a function of z. Now generalize this by slicing the n-dimensional hypersphere of radius R perpendicular to the nth axis producing a collection of parallel $(n-1)$-dimensional hyperspheres of varying radii r. Then express $V_n(R)$ as an integral of $V_{(n-1)}(r)$ with r varying as a function of the nth coordinate.

5. Looking at your results for the hypervolumes of the n-dimensional hyperspheres, deduce two general patterns for $V_n(R)$. *The formulas for n even and for n odd are different.* Explain how you got your formulas. Does your "odd" formula hold for the case $n=1$; that is, for the length of the interval $[-R,R]$?

6. Use mathematical induction to prove your two formulas for $V_n(R)$. (Use the hint from #4.) This may be hard; so get help from your instructor.

10.12 Project: Average Temperatures

A 6 inch burner on an electric stove occupies the circle $x^2 + y^2 = 9$ with $z = 0$ and is kept at the constant temperature of $120°C$. A frying pan, a skillet and a pot of water are placed on this burner. You are to plot each of these and find their average temperatures (exactly if possible and approximately to 5 decimals otherwise). Their geometries and their temperatures are given below.

Background: The average value of a function $f(x)$ on the interval $[a, b]$ is given by

$$f_{\text{ave}} = \frac{1}{b-a} \int_a^b f(x)\, dx$$

where $b - a$ is the length of the interval $[a, b]$. This can be generalized to a function $f(x, y)$ on a region R_2 in the plane or to a function $f(x, y, z)$ on a region R_3 in space:

$$f_{\text{ave}} = \frac{1}{A(R_2)} \iint_{R_2} f(x, y)\, dx\, dy \qquad f_{\text{ave}} = \frac{1}{V(R_3)} \iiint_{R_3} f(x, y, z)\, dx\, dy\, dz$$

(See Chapter 5.) Here $A(R_2)$ is the area of R_2 and $V(R_3)$ is the volume of R_3. It can also be generalized to a function $f(x, y, z)$ on a curve C parametrized by $\vec{r}(t)$ or to a function $f(x, y, z)$ on a surface S parametrized by $\vec{R}(u, v)$:

$$f_{\text{ave}} = \frac{1}{L(C)} \int_C f(\vec{r}(t))\, ds \qquad f_{\text{ave}} = \frac{1}{A(S)} \iint_S f(\vec{R}(u, v))\, dS$$

(See Chapter 6.) Here $L(C)$ is the length of C and $A(S)$ is the area of S.

For example, a cylindrical pot with dimensions $0 \le z \le 4$ and $0 \le r \le 3$ has volume

```
> V:=Muint(r, r=0..3, theta=0..2*Pi, z=0..4); V:=value(%);
```

$$V := \int_0^4 \int_0^{2\pi} \int_0^3 r\, dr\, d\theta\, dz$$

$$V := 36\,\pi$$

Suppose it is filled with water at temperature

```
> T := 120 - 3 * sqrt((r-3)^2+z^2);
```

$$T := 120 - 3\sqrt{r^2 - 6\,r + 9 + z^2}$$

You can attempt to find its average temperature exactly using

```
> Tave:=1/V * Muint('T'*r, r=0..3, theta=0..2*Pi, z=0..4);
> Tave:=value(%);
```

but Maple is unable to compute the integral. So you can find its average temperature approximately using

```
> Tave:=1/V * Muint('T'*r, r=0..3, theta=0..2*Pi, z=0..4);
> Tave:=evalf(%);
```

$$Tave := \frac{1}{36}\left(\frac{1}{\pi}\int_0^4\int_0^{2\pi}\int_0^3 T\,r\,dr\,d\theta\,dz\right)$$

$$Tave := 112.8241345$$

However, if Maple is still unable to compute the integral, then you can approximate it using

```
> n:=4:
> Tave:=1/V * middlesum( middlesum( 2*Pi*T*r, r=0..3, n), z=0..4, n):
> Tave:=evalf(%);
```

$$Tave := 112.8205678$$

Notice we have done the θ integral exactly. Be sure to increase the number of intervals until you get the desired accuracy.

1. The 8 inch frying pan occupies the circle $x^2 + y^2 \leq 16$ with $z = 0$ The temperature in polar coordinates is $T = 120 - 3|r - 3|$. (See Section 5.2.) The frying pan and the burner are shown here:

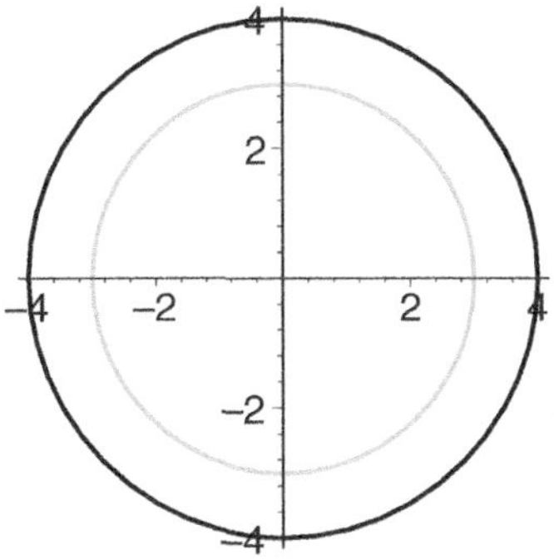

2. The 8 inch "square" skillet occupies the region inside $x^4 + y^4 = 256$. Verify this is the polar curve $r = \dfrac{4}{\sqrt[4]{\cos^4\theta + \sin^4\theta}}$. The temperature is $T = 120 - 3|r - 3|$. (You may need to split the integral for $r < 3$ and $r > 3$.)

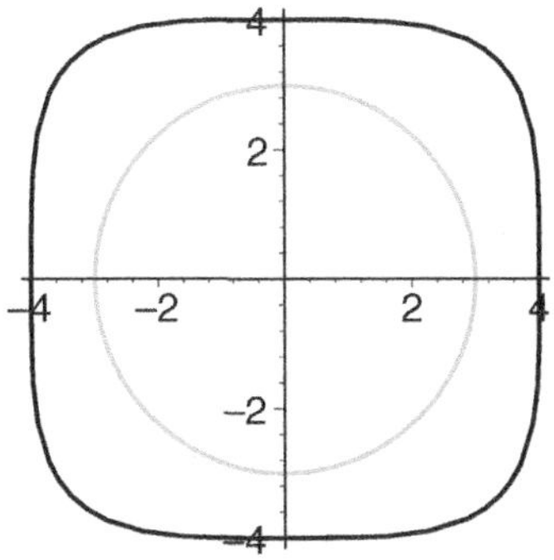

3. The water in the 8 inch pot which is 4 inches deep occupies the region given in cylindrical coordinates by $\sqrt[4]{175} - \sqrt[4]{256 - r^4} \le z \le \sqrt[4]{175}$ and has temperature given in cylindrical coordinates by $T = 120 - 3\sqrt{(r-3)^2 + z^2}$. (See Section 5.2.)

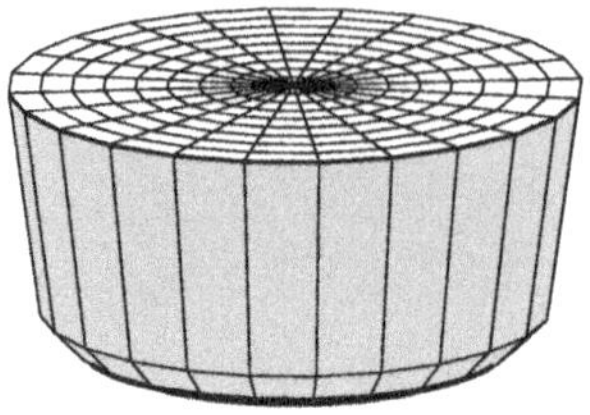

4. A temperature probe is placed in the pot of water of part 3 and measures the average temperature along a spiral curve parametrized by

$$(x, y, z) = \vec{r}(t) = (t\cos(t\pi), t\sin(t\pi), 2) \quad \text{for } 0 \le t \le 4$$

The temperature is given in cylindrical coordinates by $T = 120 - 3\sqrt{(r-3)^2 + z^2}$. (What is the percent error between the temperature measured by the probe and the actual average temperature of the water from part 3.) (See Section 6.1.)

5. The pot from part 3 occupies the surface given by $z = \sqrt[4]{175} - \sqrt[4]{256 - r^4}$ for $r \le 4$ in cylindrical coordinates and may be parametrized by

$$(x, y, z) = \vec{R}(r, \theta) = \left(r\cos(\theta), r\sin(\theta), \sqrt[4]{175} - \sqrt[4]{256 - r^4}\right) \quad \text{for } 0 \le r \le 4 \text{ and } 0 \le \theta \le 2\pi$$

The temperature is given in cylindrical coordinates by $T = 120 - 3\sqrt{(r-3)^2 + z^2}$. (See Section 6.2.)

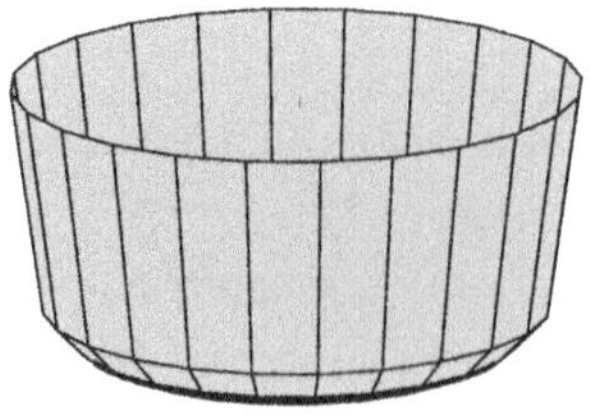

10.13 Project: The Center of Mass of Planet X

As a space pioneer, you have just arrived in a new solar system and discovered a new planet, hereafter called Planet X, which is very similar to Earth.

In order to safely orbit and land on the planet, you need to know the total mass of the planet to within $\pm 10^{22}$ kg and the center of mass of the planet to within 1 m accuracy. That is the objective of this project.

From distant but detailed radar observations, you have determined that (i) sea level is at a radius of 6371 km from the center of Planet X and that (ii) the land surface (both above and below sea level) is given in km as a radial function of the spherical coordinates (θ, ϕ) by the formula:

$$R = 6370 - 3\sin(\theta)\cos\left(2\phi + \frac{\pi}{4}\right) - 4\sin\left(2\theta + \frac{\pi}{3}\right)\cos\left(3\phi + \frac{\pi}{4}\right) + 5\cos\left(3\theta + \frac{\pi}{6}\right)\cos\left(4\phi + \frac{\pi}{3}\right)$$

You may assume that the density of water is 1 g/cm^3 or 10^{12} kg/km^3 while the average density of the land is 5.52 g/cm^3 or 5.52×10^{12} kg/km^3.

Procedure:

1. Initialize your worksheet and define the spherical coordinates:

```
> with(VecCalc):  VCalias:
> jacobian:=rho^2*sin(phi);
> x0:=rho*sin(phi)*cos(theta);
> y0:=rho*sin(phi)*sin(theta);
> z0:=rho*cos(phi);
```

   Then enter the values for the `water_density` and the `land_density` in kg/km^3 and define the `water_level` and the `land_level` in km.

2. Recreate the above plot of planet X but displayed from an orientation you prefer. Use `sphereplot` to draw two separate plots of the water surface in blue and the land surface in green. In each plot use a grid with 49 lines in the θ direction and 25 lines in the ϕ direction. This will put one line at every $7.5° = \frac{\pi}{24}$ rad. Then `display` the two plots together. (Use a courser grid until you perfect your plots.) Optional: Make the plot rotate.

3. Compute the mass and center of mass of the solid land of Planet X (not including the water).
4. Compute the mass and center of mass of the water portion of Planet X (not including the land).
5. Compute the total mass and total center of mass of Planet X by combining those for the land and water portions.

HINTS:

- Compute the integrals in spherical coordinates using `Muint` and `value` or `evalf` and remember to include the spherical Jacobian in the integrand. Then when you compute the x, y, and z coordinates of the center of mass be sure to express x, y, and z in spherical coordinates.
- Maple may not be able to compute the exact values of the triple integrals using `value` and may not even be able to compute approximate decimal values using `evalf`. In that case, you should use the midpoint rule to approximate each of the three integrals. For example, the volume of a sphere of radius 10 m can be computed from the integral

```
> Muint(1*jacobian,rho=0..10,theta=0..2*Pi,phi=0..Pi);
```

$$\int_0^{\pi}\int_0^{2\pi}\int_0^{10} \rho^2 \sin(\phi)\, d\rho\, d\theta\, d\phi$$

Then its exact and approximate values are

```
> value(%); V:=evalf(%);
```

$$\frac{4000\,\pi}{3}$$

$$V := 4188.790204$$

However, an approximate value can also be obtained from

```
> n:=8:
> middlesum( middlesum( middlesum( 1*jacobian, rho=0..10, n),
> theta=0..2*Pi, n), phi=0..Pi, n):
> V:=evalf(%);
```

$$V := 4199.358858$$

Be sure to increase the number of intervals until you get the desired accuracy.

- When you compute the mass and center of mass of the water, you must remember that (i) there is no water when the land level is above sea level and (ii) when the land level is below sea level there is only water between the land level and sea level. You can implement (ii) by taking the limits on the radial integral to be the land level and the sea level. To implement (i), you will need to use the Heaviside function which has the value 0 when its argument is negative and has the value 1 when its argument is positive. See `?Heaviside`.

10.14 Project: Steradian Measure

In this project, you will learn about steradian measure, which is a measure of solid angle, and use it to measure the solid angle subtended by several shapes.

Definition: The solid angle $\angle PS$ subtended by a smooth parametric surface S as seen from a point P (not on S) is the set of rays (half-lines) starting at P and passing through S. These rays intersect the sphere of radius R centered at P in a surface $T(R)$ with area $A(R)$. Then the steradian measure of the solid angle $\angle PS$ relative to the sphere of radius R is

$$|\angle PS| = \frac{A(R)}{R^2}$$

Thus the steradian measure is the fraction of the sphere subtended times 4π. This is analogous to the radian measure of a planar angle which is the fraction of a circle subtended times 2π .

1. Show that the steradian measure of the solid angle $\angle PS$ can be computed from the following integral over the surface S :

$$|\angle PS| = \iint_S \frac{1}{r^3}\vec{r} \cdot d\vec{S}$$

where $\vec{r}$ is the vector from P to the point on the surface S and $r = |\vec{r}|$.

HINT: Choose R so that S is completely outside the sphere of radius R. Then apply Gauss' Divergence Theorem to the solid region between S and $T(R)$ using the vector field $\vec{F} = \frac{1}{r^3}\vec{r}$.

2. Show that the steradian measure of the solid angle $\angle PS$ is independent of the radius R.

HINT: Apply Gauss' Divergence Theorem to the solid region between $T(R_1)$ and $T(R_2)$

3. Use the formula in problem #1 to compute the solid angle subtended by 3 or 4 surfaces. The following are possible surfaces but you may use any surfaces of your choice. Give appropriate plots.

   (a) The square $\{z = 3, -1 \le x \le 1, -1 \le y \le 1\}$ as seen from the origin.

   (b) The ellipse $\{z = 5, 9x^2 + 16y^2 \le 25\}$ as seen from the origin.

   (c) The paraboloid $z = x^2 + y^2 + 1$ as seen from the origin. (Think about the rays which are tangent to the paraboloid.)

   (d) The paraboloid $z = x^2 + y^2 + 1$ as seen from $(0,0,2)$.

   (e) The upper sheet of the hyperboloid of two sheets $z^2 = x^2 + y^2 + 1$ as seen from the origin.

   (f) The upper sheet of the hyperboloid of two sheets $z^2 = x^2 + y^2 + 1$ as seen from $(0,0,2)$.

   (g) The hyperboloid of one sheet $z^2 = x^2 + y^2 - 1$ as seen from the origin.

   (h) The torus given in cylindrical coordinates as $z^2 + (r-a)^2 = b^2$ as seen from the origin. (First try $a = 5$ and $b = 3$.)

Appendix A

Maple Interfaces and Resources

This chapter will discuss which Maple interface to use with this book, how to configure your worksheets and additional resources in the Standard Interface and at Maplesoft's website.

A.1 Selecting an Interface and Mode

Maple comes with two interfaces: the Standard Interface (Maple 12 on the Start menu) and the Classic Interface (Classic Worksheet Maple 12 on the Start menu). Further the Standard Interface can be used in either Worksheet Mode or Document Mode.

Whenever you start a new worksheet you need to choose an *interface* and a *mode*. Pick it and stick with it. There are incompatiblilties in going back and forth. The Standard Interface is being improved with each new edition of Maple and has many bells and whistles not available in the Classic Interface. The Classic Interface is a little faster and avoids some problems with copy and paste and with hidden characters that are present in the Standard Interface. However, the Classic Interface is not being maintained and does not have any of the new features available in the Standard Interface (which will be described later in this appendix). The Worksheet Mode has a linear structure in which you enter a Maple command on an input line and the result appears directly below it on an output line. The Document Mode is designed to produce beautiful (printable or online) documents in which you can position various elements (text, plots, tables, spreadsheets, buttons, menus, etc.) wherever you want on each page, with the Maple commands either visible or hidden.

For doing the assignments in this text, we recommend working with either the Classic Interface or the Worksheet Mode of the Standard Interface. Do not use the Document Mode; it is unnecessarily complex. For each of these interfaces, there are configuration options that we recommend. These are described below. These can be set once and for all so that you do not need to worry about them unless you go to a different computer.

A.2 The Classic Interface

In the Classic Interface, there are three options to be set for optimal performance. These are explained below. However the recommended settings are obtained by using the following sequences of menu selections:

- FILE > PREFERENCES > GENERAL > AUTOSAVE > ENABLE AUTOSAVE EVERY 3 MINUTES
- FILE > PREFERENCES > I/O DISPLAY > ASSUMED VARIABLES > NO ANNOTATION
- FILE > PREFERENCES > I/O DISPLAY > INPUT DISPLAY > MAPLE NOTATION

After making these three selections, click on Apply Globally to make the changes permanent.

AutoSave: When you are doing a long complicated computation, Maple will sometimes lock up. If you have not saved recently, you may lose a lot of work. The AutoSave feature will save most of your work. Every 3 minutes is a reasonable compromise between saving too frequently and not frequently enough. If your worksheet is named "myfile.mws", the autosaved file will be named "myfile_MAS.mws".

Assumed Variables: Maple has the ability to make assumptions about quantities and then use them. For example, if you execute

```
> assume(a>b, b>0);
```

then Maple can answer such questions as

```
> is(sqrt(a^2-b^2):: real);
```

$$true$$

However, if the ASSUMED VARIABLES option is set to TRAILING TILDES, then any formula involving variables with assumptions will have tildes following the those variables, to remind you there are assumptions. For example

```
> b^2-4*a*c;
```

$$b\tilde{}^2 - 4\, a\tilde{}\, c$$

When the formula gets long, the tildes make the formula longer and can make it hard to read. However, if the ASSUMED VARIABLES option is set to NO ANNOTATION, then the tildes are eliminated. For example

```
> b^2-4*a*c;
```

$$b^2 - 4\, a\, c$$

In this text we will make no annotation on the assumed varibles.

Input Display: Maple input can be entered either in 1-dimensional notation such as

```
> x/(x^2+1);
```

or in 2-dimensional notation such as

$$> \quad \frac{x}{x^2+1}$$

The 1-dimensional notation is called MAPLE NOTATION. The 2-dimensional notation is called STANDARD MATH NOTATION. You type exactly the same thing (`x/(x^2+1)`) to get both of these, however, the input will appear as specified by the INPUT DISPLAY option. To change this setting temporarily for one line, when the cursor is at a new prompt, select INSERT > STANDARD MATH INPUT or INSERT > MAPLE INPUT as appropriate. If you have already typed an input line and want to convert it to the other format, highlight the input with the mouse and select FORMAT > CONVERT TO > STANDARD MATH INPUT or FORMAT > CONVERT TO > MAPLE INPUT as appropriate.

The advantage of the 2-dimensional input is that it looks like standard math. The advantage of the 1-dimensional input is that it looks like what you actually typed. Further it is easier to type corrections to the Maple input than to the Standard input. (Try it!) In this book, we have typed everything in Maple notation, because it is then easier for you to know what to type. Further, we recommend that you also use Maple notation, because it is then easier for your instructor to know what you typed when grading your worksheets. However, most instructors will also accept Standard notation if you really prefer it.

Input Palettes: As an aid to figuring out what to type to get a particular input, Maple provides a set of palettes which may be used to select an input format at the click of a mouse. These can all be displayed by selecting VIEW > PALETTES > SHOW ALL PALETTES. You can click any button on the Symbol, Expression, Matrix or Vector palettes and the symbol will be inserted into the worksheet. If, for example, you click the a^b button on the Expression palette, then in Maple input you will see

```
> (%?^%?);
```

while in Standard input you will see

$$> \quad ?^?$$

Each %? or ? denotes a place where you can type any quantity you wish. You can move between input regions using ⟨TAB⟩ or ⟨SHIFT-TAB⟩. As you move through this book you may find some of these palettes useful. Use them as you wish. Experienced Maple users tend to remember the syntax.

A.3 The Standard Interface

In the Standard Interface, there are four options to be set for optimal performance. These are explained below. However the recommended settings are obtained by using the following sequences of menu selections:

- TOOLS > OPTIONS > GENERAL > AUTOSAVE > AUTOSAVE EVERY 3 MINUTES
- TOOLS > OPTIONS > DISPLAY > ASSUMED VARIABLES > NO ANNOTATION
- TOOLS > OPTIONS > INTERFACE > DEFAULT FORMAT FOR NEW WORKSHEETS > WORKSHEET
- TOOLS > OPTIONS > DISPLAY > INPUT DISPLAY > MAPLE NOTATION

After making these four selections, click on Apply Globally to make the changes permanent.

AutoSave: See the discussion above under The Classic Interface, except that if your worksheet is named "myfile.mw", the autosaved file will be named "myfile_MAS.bak" and you need to change the extension to ".mw" before you can open it.

Assumed Variables: See the discussion above under The Classic Interface.

Default Format for New Worksheets: When you open a new worksheet using FILE > NEW, you get a choice between the DOCUMENT MODE and the WORKSHEET MODE. Always choose WORKSHEET MODE. The Document allows for fancy layout on the page (which is unnecessary for this course) and is significantly more complicated to use.

When you open a new worksheet by starting Maple, by clicking on the open icon on the toolbar or by typing ⟨CTRL-N⟩, the new worksheet is opened using the mode specified by the DEFAULT FORMAT FOR NEW WORKSHEETS option. So set this to WORKSHEET.

Input Display: Maple input can be entered either in 1-dimensional notation such as

```
> x/(x^2+1);
```

or in 2-dimensional notation such as

$$> \quad \frac{x}{x^2+1}$$

The 1-dimensional notation is called MAPLE NOTATION. The 2-dimensional notation is called 2-D MATH NOTATION. You type almost the same thing to get both of these, `x/(x^2+1)` in Maple notation and `x/x^2` right-arrow `+1` in 2-D notation. Here "right-arrow" means the cursor key.

Which input will appear is specified by the INPUT DISPLAY option. To change this setting temporarily for one line, when the cursor is at a new prompt, select INSERT > 2-D MATH or INSERT > MAPLE INPUT as appropriate. If you have already typed an input line and want to convert it to the other mode, click on the input with the mouse and select FORMAT > CONVERT TO > 2-D MATH INPUT or FORMAT > CONVERT TO > 1-D MATH INPUT as appropriate.

The advantage of the 2-dimensional input is that it looks like standard math. The advantage of the 1-dimensional notation is that it looks like what you actually typed. Further it is easier to type corrections to the Maple input than to the Standard input – there are many more hidden characters. (Try it!) In this book, we have typed everything in Maple notation, because it is then easier for you to know what to type. Further, we recommend that you also use Maple notation, because it is then easier for your instructor to know what you typed when grading your worksheets. However, most instructors will also accept Standard notation if you really prefer it.

Input Palettes: As an aid to figuring out what to type to get a particular input, Maple provides a set of palettes which may be used to select an input format at the click of a mouse. The first time you open Maple, there is a list of palettes along the left side as shown below.

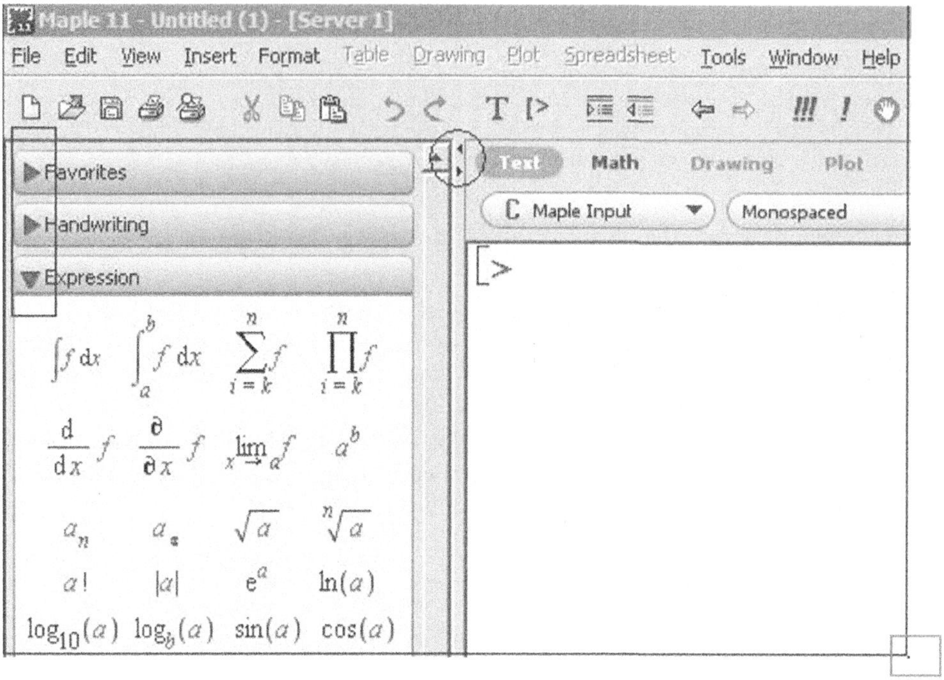

You can hide or reveal the list of palettes by clicking on the small left and right arrows shown in a circle in the above figure. You can expand or collapse each palette by clicking on the large right or down arrows shown in a rectangle in the above figure. You can click any button in any palette and the symbol will be inserted into the worksheet. If, for example, you click the a^b button on the Expression palette, then in Maple input you will see

```
> (a)^(b);
```

while in Standard input you will see

> a^b

Each a or b denotes a place where you can type any quantity you wish. You can move between input regions using TAB or SHIFT-TAB. As you move through this book you may find some of these palettes useful. Use them as you wish. Experienced Maple users tend to remember the syntax.

Tutors and Assistants: The Standard interface provides easy access to a collection of Tutors and Assistants. Just click on TOOLS ¿ TUTORS or TOOLS ¿ ASSISTANTS and select the one you want. Each of these is a pop-up window (called a maplet) which helps you through the computations on some topic. As you work through this book, periodically look at the lists of Tutors and Assistants to see if there is one on the topic you are learning.

A.4 The Maplesoft Website

There are many additional resources available at the Maplesoft website:
http://maplesoft.com/
Some are free and some are for sale.

The primary free resource is the Maple Application Center at:
http://maplesoft.com/applications/
It offers numerous Maple worksheets, packages and maplets in such diverse areas as mathematics, physics, chemistry and engineering including the `VecCalc` package by Belmonte and Yasskin for multivariable vector calculus.

The commercial resources are available at the Maple Connect site:
http://www.maplesoft.com/products/thirdparty/
including the Maplets for Calculus. Maplesoft also has a site dedicated to students:
http://maplesoft.com/students/
On this page, we point out the Maple 12 Calculus Kit which includes a copy of Maple 12, the Precalculus Study Guide, the Calculus Study Guide and the Maplets for Calculus. The Calculus Study Guide is a collection of Maple worksheets on various topics in single variable calculus. The Maplets for Calculus, which are described in detail at:
http://m4c.math.tamu.edu/
is a collection of maplets (Maple applets) which teach various topics in calculus whereas the built-in Tutors simply do the computations for you. If you already own a copy of Maple 12 (e.g., it was packaged with your copy of your Calculus textbook), then the other parts are available individually.

Appendix B

The VecCalc Package

By Philip B. Yasskin and Arthur Belmonte
Department of Mathematics, Texas A&M University

B.1 Description of the Package

The `VecCalc` package is a collection of *Maple* commands designed for the study and application of vector calculus problems.

The current version of `VecCalc` is version 8.0 which works with *Maple* 7 through *Maple* 12, except that a bug in Maple 9.5 causes Vectors of functions to display incorrectly (although they still operate correctly). There are also versions of `VecCalc` for earlier versions of *Maple* but they are not being maintained. The `VecCalc` version is displayed when you start the package. Current information about the package is available at either:

- `http://calclab.math.tamu.edu/maple/veccalc/`
- `http://maplesoft.com/applications/` Keyword: `VecCalc`

All of this book was checked with `VecCalc` 8.0 in *Maple* 10 through 12 using both the Classic Interface and the Worksheet Mode of the Standard Interface.

B.2 Acknowledgments and History

The `VecCalc` commands were originally written for *Maple* V Release 3 by A. Belmonte and P. Yasskin. The commands were organized into a package called `vec_calc` for *Maple* V Release 3 by James Warren and P. Yasskin. The help pages were first written for *Maple* V Release 3 by David Arnold, J. Warren and P. Yasskin and converted to *Maple* V Release 4 by Ken Parker, Jared Teslow and P. Yasskin. The package was converted to a module and the commands were updated to *Maple* 6 and 7 by A. Belmonte and the help pages were updated to *Maple* 6 and 7 by P. Yasskin. The module, commands and help pages were significantly modified for *Maple* 8 to be compatible with new Vector and Matrix types and renamed `VecCalc` by P. Yasskin with help from Chad Wellington, Krista Rister, Chris Haag, Ethan McConnell, Allison DenBleyker and Jeffrey Yasskin.

B.3 Obtaining and Installing the Files

Before using the package, you must first obtain and install three files:

File	Filename	Size
package index	`VecCalc.ind`	40 KB
package library	`VecCalc.lib`	42 KB
package help	`VecCalc.hdb`	545 KB

These files are exactly the same for all operating systems. They are available at:

- `http://calclab.math.tamu.edu/maple/veccalc/`

Save the files on your hard disk. Here are the recommended locations:

- Windows: `C:\Program Files\Maple 12\lib`
- Unix/Linux: `/usr/local/maple12/lib`
- Macintosh OS X: `/Library/Frameworks/Maple.framework/Versions/12/lib`

If you put the files in a different directory or put them on a CD or USB drive, then the instructions below must be appropriately modified.

B.4 Using the Package

To use the commands in the `VecCalc` package, you must first execute up to three commands. The first command tells *Maple* where the package library files are located. If you saved the files in the recommended location listed above, then this command is unnecessary. Otherwise, after starting *Maple*, execute the command

```
> libname := libname, "yourdirectorypath";
```

where `yourdirectorypath` is the full path to the directory where you saved the `VecCalc` files using / between directories. For example, on Windows, if you saved them on a CD or USB drive in drive E:, then the command will be

```
> libname := libname, "E:/";
```

and if you save them in the directory `D:\mypath\mydir` then the command will be

```
> libname := libname, "D:/mypath/mydir";
```

NOTE: The path is enclosed in **double quotes** (").

The second command reads in the package commands:

```
> with(VecCalc);
```

Finally, the third (optional) command defines many abbreviations for the `VecCalc` commands:

```
> VCalias;
```

This book assumes that you have executed this optional command so that all the aliases are available.

Below is the output you should expect from these commands.

```
> libname := libname, "E:/";
```

$$libname := \text{"C:\textbackslash Program Files \textbackslash Maple 12/lib"}, \text{"E:/"}$$

```
> with(VecCalc);

Initializing Package:   VecCalc   Version 8.0

All HELP is linked from ?VecCalc

To use aliases, execute:   VCalias;

Loading packages:   student   plots
```

[&- >, &., &@, &x, *Angle*, *Cross*, *Curl*, *CurveAcceleration*, *CurveArcLength*, *CurveBinormal*, *CurveCurvature*, *CurveForget*, *CurveJerk*, *CurveNormal*, *CurveNormalAcceleration*, *CurveSpeed*, *CurveTangent*, *CurveTangentialAcceleration*, *CurveTorsion*, *CurveVelocity*, D, *Determinant*, *Diff*, *Divergence*, *Dot*, *Gradient*, *Hessian*, *Int*, *JacobianDeterminant*, *JacobianMatrix*, *Laplacian*, *LeadingPrincipalMinorDeterminants*, *Length*, *Limit*, *LineIntScalar*, *LineIntScalarInert*, *LineIntVector*, *LineIntVectorInert*, *MakeFunction*, *Multipleint*, *ScalarPotential*, *SurfaceArea*, *SurfaceForget*, *SurfaceIntScalar*, *SurfaceIntScalarInert*, *SurfaceIntVector*, *SurfaceIntVectorInert*, *SurfaceNormal*, *SurfaceNormalLength*, *SurfaceTangents*, *VCalias*, *VectorPotential*, *cyl2rect*, *cyl2sph*, *deg2rad*, *diff*, *evalFunction*, *init*, *int*, *limit*, *multipleint*, *polar2rect*, *rad2deg*, *rect2cyl*, *rect2polar*, *rect2sph*, *simplifyvec*, *sph2cyl*, *sph2rect*, *value*]

```
> VCalias;
```

Point, *MF*, *EF*, *Det*, *LPMD*, *d2r*, *r2d*, *p2r*, *r2p*, *c2r*, *r2c*, *s2r*, *r2s*, *s2c*, *c2s*, *Cv*, *Ca*, *Cj*, *Cs*, *CL*, *CT*, *CN*, *CB*, *Ck*, *Ct*, *CaT*, *CaN*, *Cforget*, *ST*, *SN*, *SNL*, *SA*, *Sforget*, *Grad*, *Hess*, *Div*, *Lap*, *Jac*, *JDet*, *SPot*, *VPot*, *Muint*, *muint*, *Lis*, *lis*, *Liv*, *liv*, *Sis*, *sis*, *Siv*, *siv*

After starting the `VecCalc` package, you may get help on any command by executing

```
> ?VecCalc
```

and following the hyperlinks.

Default Output Types: The `VecCalc` package can work either with `Vector` and `Matrix` types or with `list` types. This book uses `Vector[row]` and `Matrix`, but some commands are easier with `list`. You are free to use either type for input at any time. Normally the output type will match the input type. However, a few commands need to be told which type to use for output. To change the default output type, execute:

```
> OutputVectorType:=list:  OutputMatrixType:=list:
```

The other choices for `OutputVectorType` are `Vector[row]` (the initial default) or `Vector[column]`. The other choice for `OutputMatrixType` is `Matrix` (the initial default).

Appendix C

Tables of Applications of Integration

This appendix provides three tables of applications of integration. The first contains applications of double and triple integrals. The second contains applications of line and surface integrals of scalar fields. The third contains applications of line and surface integrals of vector fields with alternate forms due to the Fundamental Theorem of Calculus for Curves, Stokes' Theorem and Gauss' Theorem, when appropriate. Examples appear throughout the text.

Table C.1: Table of Applications of Multiple Integrals

Application[a]	**2-D**	**3-D**
differential	$dA = dx\,dy = r\,dr\,d\theta = J\,du\,dv$	$dV = dx\,dy\,dz = r\,dr\,d\theta\,dz = \rho^2\sin(\phi)\,d\rho\,d\theta\,d\phi = J\,du\,dv\,dw$
measure	$A = \iint_R 1\,dA$ area	$V = \iiint_R 1\,dV$ volume
total mass[b]	$M = \iint_R \rho\,dA$	$M = \iiint_R \rho\,dV$
electric charge [c]	$Q = \iint_R \rho_c\,dA$	$Q = \iiint_R \rho_c\,dV$
moments	$M_y = \iint_R x\,\rho\,dA$ $M_x = \iint_R y\,\rho\,dA$	$M_{yz} = \iiint_R x\,\rho\,dV$ $M_{xz} = \iiint_R y\,\rho\,dV$ $M_{xy} = \iiint_R z\,\rho\,dV$
center of mass [d]	$(\overline{x}, \overline{y}) = \left(\frac{M_y}{M}, \frac{M_x}{M}\right)$	$(\overline{x}, \overline{y}, \overline{z}) = \left(\frac{M_{yz}}{M}, \frac{M_{xz}}{M}, \frac{M_{xy}}{M}\right)$
moments of inertia [e]	$I_x = \iint_R y^2\,\rho\,dA$ $I_y = \iint_R x^2\,\rho\,dA$ $I_0 = \iint_R (x^2 + y^2)\,\rho\,dA$	$I_x = \iiint_R (y^2 + z^2)\,\rho\,dV$ $I_y = \iiint_R (x^2 + z^2)\,\rho\,dV$ $I_z = \iiint_R (x^2 + y^2)\,\rho\,dV$
radii of gyration	$\overline{\overline{x}} = \sqrt{\frac{I_y}{M}}$, $\overline{\overline{y}} = \sqrt{\frac{I_x}{M}}$	N/A

[a] R = region of integration
[b] ρ = mass density
[c] ρ_c = charge density
[d] The center of mass is also called the centroid when the density is a constant.
[e] Note that $I_0 = I_x + I_y$.

Table C.2: Table of Applications of Line and Surface Integrals of Scalars

Application	**Line Integrals**[a]	**Surface Integrals**[b]
scalar differential	$ds = \lvert\vec{v}\rvert\, dt$	$dS = \lvert\vec{N}\rvert\, du\, dv$
measure	$L = \int_A^B 1\, ds$ arc length	$A = \iint_S 1\, dS$ surface area
total mass [c]	$M = \int_A^B \rho\, ds$	$M = \iint_S \rho\, dS$
electric charge [d]	$Q = \int_A^B \rho_c\, ds$	$Q = \iint_S \rho_c\, dS$
moments	$M_{yz} = \int_A^B x\,\rho\, ds$ $M_{xz} = \int_A^B y\,\rho\, ds$ $M_{xy} = \int_A^B z\,\rho\, ds$	$M_{yz} = \iint_S x\,\rho\, dS$ $M_{xz} = \iint_S y\,\rho\, dS$ $M_{xy} = \iint_S z\,\rho\, dS$
center of mass [e]	$(\overline{x}, \overline{y}, \overline{z}) = \left(\frac{M_{yz}}{M}, \frac{M_{xz}}{M}, \frac{M_{xy}}{M}\right)$	$(\overline{x}, \overline{y}, \overline{z}) = \left(\frac{M_{yz}}{M}, \frac{M_{xz}}{M}, \frac{M_{xy}}{M}\right)$
moments of inertia	$I_x = \int_A^B (y^2 + z^2)\,\rho\, ds$ $I_y = \int_A^B (x^2 + z^2)\,\rho\, ds$ $I_z = \int_A^B (x^2 + y^2)\,\rho\, ds$	$I_x = \iint_S (y^2 + z^2)\,\rho\, dS$ $I_y = \iint_S (x^2 + z^2)\,\rho\, dS$ $I_z = \iint_S (x^2 + y^2)\,\rho\, dS$

[a] $\vec{r}(t)$ is the curve where $A = \vec{r}(a)$ and $B = \vec{r}(b)$
[b] $S(u, v)$ is the surface
[c] ρ = mass density
[d] ρ_c = charge density
[e] The center of mass is also called the centroid when the density is a constant.

Table C.3: Table of Applications of Line and Surface Integrals of Vectors

	Line Integrals	**Surface Integrals**
vector differential	$\vec{ds} = \vec{v}\,dt = \hat{T}\,ds$	$\vec{dS} = \vec{N}\,du\,dv = \hat{N}\,dS$
open curve[a]/surface[b]	$\mathcal{W}ork = \int_A^B \vec{F}\cdot\vec{ds}$ $= f(B) - f(A)$ if $\vec{F} = \vec{\nabla} f$ by Fund. Thm. of Calc. for Curves	$\mathcal{F}lux = \iint_S \vec{F}\cdot\vec{dS}$ $= \oint_{\partial S} \vec{A}\cdot\vec{ds}$ if $\vec{F} = \vec{\nabla}\times\vec{A}$ by Stokes' Theorem
closed curve[c]/surface[d]	$\mathcal{C}irculation = \oint_{\vec{r}} \vec{F}\cdot\vec{ds}$ $= \iint_S \vec{\nabla}\times\vec{F}\cdot\vec{dS}$ by Stokes' Theorem where $\partial S = \vec{r}$	$\mathcal{E}xpansion = \iint_S \vec{F}\cdot\vec{dS}$ $= \iiint_V \vec{\nabla}\cdot\vec{F}\,dV$ by Gauss' Theorem where $\partial V = S$

[a] $\vec{r}(t)$ is the open curve where $A = \vec{r}(a)$ and $B = \vec{r}(b)$
[b] $S(u,v)$ is the open surface
[c] $\vec{r}$ is the closed curve
[d] S is the closed surface

Appendix D

Programming with Maple

We have indicated in earlier chapters that Maple has a built-in programming language and have already seen a simple `do` loop. In fact, it is possible to use Maple to create highly structured programs that can perform a large number of useful tasks. Most importantly, any program written in Maple can call any of the high level commands available in Maple, or any program that you have previously written in Maple and included in the same session.

This chapter is intended as a short introduction to the Maple programming environment. This includes conditional and logical structures

```
if...then...elif/else...end if, and, or, not, true, false,
```

looping structures

```
for/while...do...end do, next, break
```

and procedural structures

```
proc...end proc, args, nargs, return, module.
```

D.1 Conditional and Logical structures

The first class of programming structures are conditional statements which tell Maple to perform certain commands provided that a certain condition is satisfied.

Example: D.1. Determine if the quadratic polynomial $3x^2 - 2x + 4$ has 2 distinct real roots.

SOLUTION: We know that ax^2+bx+c has 2 distinct real roots if the discriminant b^2-4ac is positive. To compute this, we enter a polynomial:

```
> p:=3*x^2-2*x+4;
```

$$p := 3\,x^2 - 2\,x + 4$$

read off the coefficients:

```
> a,b,c:=coeff(p,x,2), coeff(p,x,1), coeff(p,x,0);
```

$$a,\ b,\ c := 3,\ -2,\ 4$$

and test the discriminant:

```
> if b^2-4*a*c > 0
> then "2 real roots"
> else "not 2 real roots"
> end if;
```

$$\text{“not 2 real roots”}$$

So the polynomial does not have 2 real roots.

Notice we entered the `if...then...else...end if` on multiple lines. You can do this by pressing ⟨SHIFT-ENTER⟩. If you just press ⟨ENTER⟩, the commands will work but you get a warning message that the input is incomplete. Putting the `then` and `else` clauses on separate lines is good style, but is not necessary.

In this example, we could have just computed the discriminant:

```
> b^2-4*a*c;
```

$$-44$$

to see that it is negative. However, if this code is part of a larger computation and Maple needs to do something else based on the result of this test, then there might not be a human to decide if the discriminant is positive or negative.

"If" statements may be nested or additional conditions may be included by adding `elif` clauses. (This is an abbreviation for "else if.")

Example: D.2. Determine if the quadratic polynomial $x^2 - 4x + 4$ has 2 real roots, 1 real root or 2 complex roots:

SOLUTION: We add an `elif` clause to our `if` statement:

```
> q:=x^2-4*x+4;
```

$$q := x^2 - 4\,x + 4$$

```
> a,b,c:=coeff(q,x,2), coeff(q,x,1), coeff(q,x,0);
```

$$a,\ b,\ c := 1,\ -4,\ 4$$

```
> if b^2-4*a*c > 0
> then "2 real roots"
> elif b^2-4*a*c = 0
> then "1 real root"
> else "2 complex roots"
> end if;
```

$$\text{“1 real root”}$$

Repeat this with the polynomials `q:=x^2-3*x+4:` and `q:=x^2-5*x+4:` to see the difference in output.

In general, the syntax for an "if" statement is:

```
if logical condition
   then statements
   elif logical condition
     then statements
   else statements
end if;
```

Here, the `elif...then...` and `else...` clauses are optional and there may be an arbitrary number of `elif...then...` clauses. The *statements* following `then` or `else` can be any valid Maple commands, with each statement ending in either a colon or semicolon. The *logical conditions* following `if` or `elif` must evaluate to `true` or `false` and may be constructed using, for instance, the math and set relations:

Relation	**Math Symbol**	**Maple Symbol**
equal	$=$	`=`
not equal	$\neq$	`<>`
less than	$<$	`<`
less than or equal	$\leq$	`<=`
greater than	$>$	`>`
greater than or equal	$\geq$	`>=`
subset	$\subseteq$	`subset`
element of	$\in$	`in`

as well as the logical and set operators:

Operator	**Math Symbol**	**Maple Symbol**
and	$\wedge$	`and`
or	$\vee$	`or`
exclusive or		`xor`
not	$\neg$	`not`
implies	$\Rightarrow$	`implies`
union	$\cup$	`union`
intersection	$\cap$	`intersection`
minus	$\setminus$	`minus`

The order of precedence for these operators may be found by typing `?precedence`, but to be safe *use lots of parentheses*. You can also use an explicit `true` or `false`.

Example: D.3. Given three numbers a, b and c, put them in ascending order and rename them x, y and z. Use two `if` statements: the first to put a and b in order and the second to put c in the correct position. Try it on $a = 7$, $b = -2$ and $c = 4$.

SOLUTION: Use intermediate variables m and n to denote a and b in ascending order:

```
> a,b,c:=7,-2,4;
```

$$a, b, c := 7, -2, 4$$

```
> if a<b
> then m,n:=a,b;
> else m,n:=b,a;
> end if;
```

$$m, n := -2, 7$$

```
> if c<m
> then x,y,z:=c,m,n;
> elif c<n
> then x,y,z:=m,c,n;
> else x,y,z:=m,n,c;
> end if;
```

$$x, y, z := -2, 4, 7$$

Now try it with other choices of a, b and c.

D.2 Looping Structures

The second class of programming structures are looping structures (`do` loops) which tell Maple to perform certain commands repeatedly until a certain condition is satisfied.

In general, the syntax for a "do" loop is:

```
for index from start by step to finish
  while logical condition
  do
    statements
end do;
```

Here, the *statements* between `do` and `end do` are executed repeatedly, with the variable *index* initially having the value *start* and incrementing each time by the amount *step* until either *index* becomes larger than the value *finish* or the *logical condition* in the `while` clause becomes false. The `for...`, `from...`, `by...`, `to...` and `while...` clauses are all optional. If the `from...` or `by...` clauses are missing, then the values of *start* and *step* are taken to have the default value `1`, respectively. If the `to` and `while` clauses are both missing then the loop will be repeated forever or until a `break` command is executed (as described below). The *logical condition* in the `while` clause must evaluate to `true` or `false` and is constructed exactly like those in an `if` statement. The *statements* between `do` and `end do` can be any valid Maple commands, with each statement ending in either a colon or semicolon.

Example: D.4. Find the first 4 derivatives of $f := \sin(x)\cos(x)$ and see if you can find a pattern.

SOLUTION: Define the function and use a `do` loop to compute the derivatives:

```
> f:=sin(x)*cos(x);
```

$$f := \sin(x)\cos(x)$$

```
> for n to 4 do
> diff(f,x$n)
> end do;
```

$$\cos(x)^2 - \sin(x)^2$$
$$-4\sin(x)\cos(x)$$
$$-4\cos(x)^2 + 4\sin(x)^2$$
$$16\sin(x)\cos(x)$$

NOTE: We entered multiple lines by pressing ⟨SHIFT-ENTER⟩. Further, we did not type a `from` or `by` clause; so by default, the index `n` starts at 1 and steps by 1 on each iteration.

Notice that $f^{(2)} = -4f$ and $f^{(3)} = -4f^{(1)}$. So if n is even, then $n = 2k$ and $f^{(2k)} = (-4)^k f$, while if n is odd, then $n = 2k+1$ and $f^{(2k+1)} = (-4)^k f^{(1)}$.

Finally, there are two other commands which are used to control the flow in `do` loops: `next` and `break`. The `next` command causes Maple to skip the rest of that iteration and go on to the *next* iteration. The `break` command causes Maple to *break* out of the loop by skipping the rest of that iteration and all remaining iterations. Here is an example:

Example: D.5. Find the first 10 derivatives of $f := \sin(x)\cos(x)$ and evalute them at $x = 0$. If a derivative at $x = 0$ is positive, find its value at $x = \pi/2$. If it is zero, go on to the next derivative. If it is negative, don't bother to compute the remaining derivatives.

SOLUTION:

```
> for n to 10 do
> derf[n]:=diff(f,x$n);
> derfz[n]:=eval(subs(x=0,derf[n]));
> if derfz[n]<0
> then break
> elif derfz[n]=0
> then next
> end if;
> derfp[n]:=eval(subs(x=Pi/2,derf[n]))
> end do;
```

$$derf_1 := \cos(x)^2 - \sin(x)^2$$
$$derfz_1 := 1$$
$$derfp_1 := -1$$
$$derf_2 := -4\sin(x)\cos(x)$$
$$derfz_2 := 0$$
$$derf_3 := -4\cos(x)^2 + 4\sin(x)^2$$
$$derfz_3 := -4$$

Notice the first derivative at $x = 0$ is positive, so Maple finds the first derivative at $x = \pi/2$. The second derivative at $x = 0$ is zero, so Maple skips the second derivative at $x = \pi/2$. The third derivative at $x = 0$ is negative, so Maple skips the remaining derivatives.

D.3 Procedures

Once you have written some code, you may want to reuse it with different input data. You don't want to have to retype the code every time with slight changes to the variable names or values. So you can write your own Maple commands with variable input (called parameters or arguments). These are called Maple programs or procedures. For example a procedure to add a first number to 3 times a second number is defined by

```
> mysum:=proc(a,b)
> a+3*b;
> end proc;
```

$$mysum := \mathbf{proc}(a,\ b)\, a + 3 * b\ \ \mathbf{end\ proc}$$

It is called with `a=2` and `b=1` by executing

```
> mysum(2,1);
```

$$5$$

The arguments in the call can be expressions and are totally unrelated to the names used for the parameters in the definition of the procedure. For example:

```
> mysum(b,x+y);
```

$$b + 3\,x + 3\,y$$

Notice that `a` in the definition has been replaced by `b` and `b` has been replaced by `x+y`.

More generally, the syntax for a procedure is:

```
procname:=proc(parameters)
   local variables;
   global variables;
   statements
end proc;
```

which is called by executing:

procname(*arguments*);

Here *procname* is the name you assign to the procedure. When you call the procedure, the values of the *arguments* in the call become the values of the *parameters* in the definition and then the *statements* are executed in order. The value returned by the procedure is *normally* the result of the last statement in the definition. (However, see the discussion of the `return` command below.) This returned value is the only thing printed out by the procedure, unless there are explicit `print` statements.

The `local` and `global` statements are optional. The *variables* in the `local` statement are only known within the procedure while the *variables* in the `global` statement are known both inside and outside the procedure. If you don't include these statements, Maple has complicated rules for determining if the variables are local or global. So it is best to include the statements, so that you are sure Maple is doing what you expect. The remaining *statements* in the procedure definition can be any valid Maple commands, with each statement ending in either a colon or semicolon.

Now let's create some procedures that are slightly more complex than the simple example given above. The first of these illustrates a simple use of the `for/do` looping structure

Example: D.6. In the previous section, we wrote a `do` loop to compute 4 derivatives of a function. Convert this into a procedure which will work for any function (written as an expression) and any number of derivatives. Apply the procedure to find the first 3 derivatives of the function $e^x \sin(x)$.

SOLUTION: First define the procedure

```
> derivs:=proc(f,k)
> local n;
> for n to k do
> print(diff(f,x$n))
> end do;
> end proc;
```

$$derivs := \textbf{proc}(f,\ k)\ \textbf{local}\ n;\ \textbf{for}\ n\ \textbf{to}\ k\ \textbf{do}\ \mathrm{print}(\mathrm{diff}(f,\ x\,\$\,n))\ \textbf{end do end proc}$$

NOTES: We used a local variable `n` in the procedure so that we do not mess up any variable `n` used outside the procedure. Also, we explicitly typed `print`, since otherwise Maple would not print out intermediate results.

Now apply this program to the function $e^x \sin(x)$ and ask for 3 derivatives:

```
> derivs(exp(x)*sin(x),3);
```

$$e^x \sin(x) + e^x \cos(x)$$

$$2\,e^x \cos(x)$$

$$2\,e^x \cos(x) - 2\,e^x \sin(x)$$

Example: D.7. Write a procedure that will plot a function along with its tangent lines at several points. The inputs to the procedure should be the function, the two endpoints for the plot interval, and a list of the x-coordinates of the points of tangency. Use this program to plot $4\sin(x) + \cos(2x)$ along with its tangent lines at $x = -1.5,\ 2,\ 4$ on the interval $[-3, 6]$.

SOLUTION: The procedure (discussed below) is

```
> tan_plot:=proc(f,a,b,xs)
> local plist,x,i,xi;
> plist:=f(x);
> for i from 1 to nops(xs) do
> xi:=xs[i];
> plist:=plist,D(f)(xi)*(x-xi)+f(xi);
> end do;
> plot({plist},x=a..b);
> end proc:
```

and we apply it as follows:

```
> f:=x-> 4*sin(x)+cos(2*x):
```

```
> tan_plot(f,-3,6,[-1.5,2,4]);
```

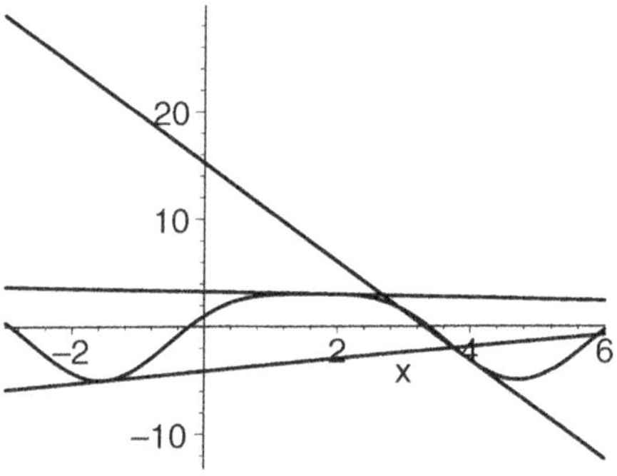

We need to discuss the details of the procedure. First, since the procedure refers to `f(x)` and `D(f)`, we must define the function using arrow notation, not as an expression. Second, we used a new command `nops` (which stands for "number of operands"). This command returns the number of entries in a list (or, more generally,the number of operands in any operation). Consequently, the loop index `i` runs from 1 to the number of x's. The variable `xi` is defined as the ith x so we don't have to recompute it three times. (As a rule of thumb, if a quantity is recomputed more than twice, give it a name.) The variable `plist` stores the functions to be plotted. Initially `plist` is just the original function. The loop then appends each of the tangent line functions and the `plot` command encloses it in braces. Further, we defined all the variables to be local except for the parameters of the procedure. Finally, note that, although there are several executable statements with semicolons, we only see the plot output from the last one. That is because a Maple procedure only returns the value of the last executable statement.

One other fact about Maple procedures is that you do not need to have a fixed number of arguments in the procedure. This is demonstrated in the next example.

Example: D.8. Rewrite the previous procedure so that the inputs are the function, the x-range and y-range for the plot, and the x-coordinates of the points of tangency. Use it to plot $x^3 - 3x^2$ along with its tangent lines at $x = 0,\ 1,\ 2,\ 3$ on the interval $[-1, 4]$.

NOTE: The x-coordinates are *not* put in a list; so you don't know in advance how many arguments there will be. Maple handles this with two new commands: `args` and `nargs`. Within the procedure, `nargs` is the number of arguments and `args` is the list of arguments.

SOLUTION: The procedure (discussed below) is

```
> tan_plot:=proc(f,xrange,yrange)
> local plist,x,i,xi;
> plist:=f(x);
> for i from 4 to nargs do
> xi:=args[i];
> plist:=plist,D(f)(xi)*(x-xi)+f(xi);
> end do;
> plot({plist},x=xrange,y=yrange);
> end proc:
```

and we apply it as follows:

```
> g:=x->x^3-3*x^2:
> tan_plot(g,-1..4,-10..10,0,1,2,3);
```

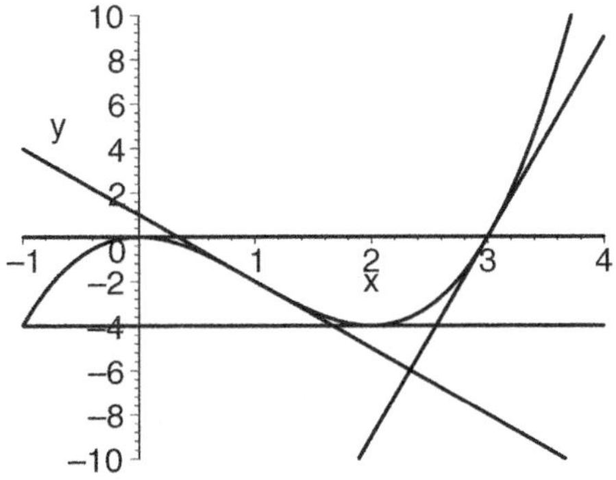

Again we need to discuss the details of the procedure. First, notice that when there are an indeterminate number of arguments, the first few are named in the procedure definition and the others remain unnamed. Second, the plot intervals are now given as ranges; so they must be entered in the form `a..b` instead of `a,b` which would count as two separate arguments. Third, instead of using `nops(xs)` we use `nargs` and instead of using `xs[i]` we use `args[i]`. Finally, the index `i` starts at 4, counting from after the three named parameters `f, xrange` and `yrange`.

Our final generalization of this example demonstrates the use of the `return` command to exit from the middle of a procedure and *return* a value which is not the result of the last statement in the procedure definition.

Example: D.9. Rewrite the previous procedure to check to see if any of the x-coordinates yield a vertical asymptote or vertical tangent. (This is a point $x = a$ where $\lim\limits_{x \to a} f'(x) = \pm\infty$.) If there is a vertical asymptote or vertical tangent, print out an error message and return that value of x. Use the procedure to plot $\dfrac{x}{x-2}$ along with its tangent lines at $x = 1,\ 2,\ 3$ on the interval $[0, 4]$. If there is a vertical asymptote or vertical tangent, drop that x-coordinate and replot.

Solution: The procedure (discussed below) is

```
> tan_plot:=proc(f,xrange,yrange)
> local plist,x,i,xi;
> plist:=f(x);
> for i from 4 to nargs do
> xi:=args[i];
> if limit(abs(D(f)(x)),x=xi)=infinity
> then print("Error:  The following point is a vertical\nasymptote
> or vertical tangent.");
> return xi;
> end if;
> plist:=plist,D(f)(xi)*(x-xi)+f(xi);
> end do;
> plot({plist}, x=xrange, y=yrange, discont=true);
> end proc:
```

and we apply it as follows:

```
> h:=x->x/(x-2);
```

$$h := x \to \frac{x}{x-2}$$

```
> tan_plot(h,0..4,-10..10,1,2,3);
```

"Error: The following point is a vertical asymptote or vertical tangent."

2

Notice how Maple exited without completing the do loop or the remaining statements in the procedure. It did not print out the result of the last statement which here is the plot. Rather it returned the value in the return statement, 2. Deleting 2 from the list of points, we get:

```
> tan_plot(h,0..4,-10..10,1,3);
```

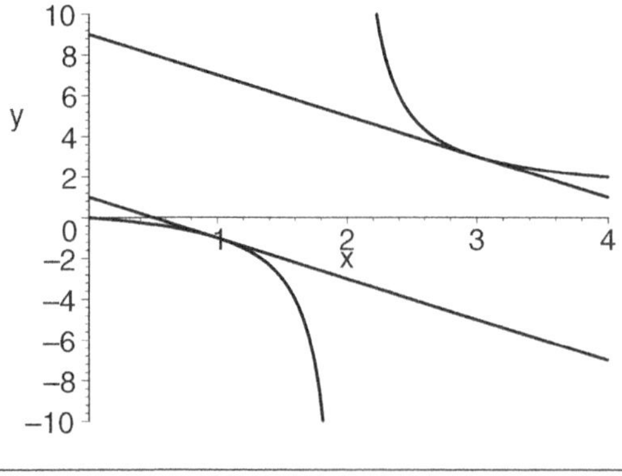

You may be surprised to realize that you have already been writing simple Maple procedures in earlier chapters. These are the functions defined using arrow notation. You may be puzzled because they do not seem to have the form given above. However, the function defined by

```
> f:=x->x^2;
```

$$f := x \to x^2$$

is actually a shorthand for the procedure

```
> f:=proc(x) x^2 end proc;
```

$$f := \textbf{proc}(x)\, x^2\, \textbf{end proc}$$

The arrow symbol tells Maple to create a procedure that has the single argument x and gives the output x^2. Both produce the exact same output when you evaluate them e.g.

```
> f(2); f(a+1);
```

$$4$$
$$(a+1)^2$$

We have given only a brief glimpse of the programming capability of Maple. We encourage you to explore all of Maple's programming capability by clicking on the HELP menu and INTRODUCTION entry and selecting `Programming...` from the first column of the Help Browser. Finally, we point out that once you have created a collection of related procedures, you can group them in a package by using the module command. See `?module` for more details. This is in fact how the `VecCalc` package was written.

D.4 Exercises

1. Enter the functions $f = \dfrac{x^2}{x + \sin(x)}$ and $g = \dfrac{x - \sin(x)}{x^2}$ using arrow notation. Use an `if` statement to determine whether $f(2) > g(2)$, $f(2) = g(2)$ or $f(2) < g(2)$.

2. Write a `do` loop which will compute $\sin(1.0)$, $\sin(\sin(1.0))$, ..., $\sin(\sin(\cdots(\sin(1.0))\cdots))$ with up to 20 applications of sin. What do you think is the limit as the number of sin's becomes infinite?

   HINT: Before the loop, execute `x:=1.0`. Then on each iteration execute `x:=sin(%)`.

   NOTE: The final answer can also be obtained from `(sin@@20)(1.0)`.

3. Write a procedure which tests a quadratic polynomial to see if it has 2 real roots, 1 real root or 2 complex roots. Test it on the polynomials $x^2 - 3x + 4$, $x^2 - 4x + 4$ and $x^2 - 5x + 4$.

4. Write a procedure which puts 4 numbers in descending order. Test the procedure on several sets of numbers.

5. Write a procedure which computes the first n derivatives of a function given in arrow notation. Apply the procedure to compute the first 4 derivatives of $\dfrac{e^x - e^{-x}}{e^x + e^{-x}}$.

6. Write a procedure which computes the first n derivatives of a given function at a given number $x = a$, but quits if any of these derivatives are zero. Use either functions or expressions. If you use expressions, remember that the result of a `subs` command is not automatically simplified. So `simplify` it. Apply the procedure to compute the first 5 derivatives of $\frac{\cos(x)}{\sec(x)+\tan(x)}$ at $x = 0$ and then at $x = \pi/4$.

Index

Made in the USA
Middletown, DE
04 November 2026

36537963R00150